Tackle Car Maintenance

By the same author
Save Money On Your Car

Stuart Bladon

Tackle
CAR MAINTENANCE

Stanley Paul · London

Stanley Paul & Co Ltd
3 Fitzroy Square, London W1
An Imprint of the Hutchinson Publishing Group

London Melbourne Sydney Auckland
Wellington Johannesburg Cape Town
and agencies throughout the world

First published 1975
© Stuart Bladon 1975
Drawings © Stanley Paul & Co Ltd 1975

Printed in Great Britain
by Flarepath Printers Ltd, St Albans, Herts,
and bound by Wm Brendon & Son Ltd of
Tiptree, Essex

ISBN 0 09 122670 8 (cased)
      0 09 122671 6 (paper)

Contents

1 · Efficient, Systematic Work

AT the outset it should be emphasized that this is a guide to planned maintenance of a car; it is not a repair manual. Sometimes the difference between maintenance procedures and those which are strictly in the category of repairs becomes a little hard to define. For example, distributor points become eroded in extended use, and have to be replaced periodically, as do sparking plugs; so how does this compare with the progressive wearing out and eventual need for replacement of the clutch, or even the engine?

The more important question is whether the work involved is within scope of the ordinary motorist having no special knowledge of the car, but anxious to reduce his motoring costs by tackling the routine maintenance himself. In the explanations of service procedures it has been assumed that the user may be a beginner, and others with a basic knowledge of the working principles of a car are asked to forgive and skip over what may seem elementary explanations of basic functions and still find helpful information in the recommended procedures.

Jobs you can't do

As important as knowing what you can do for the sound upkeep of a car is the appreciation that, mainly because of need for special equipment, many jobs are beyond the average home mechanic especially when we move into the realm of repairs. By studying the procedure described, most motorists should find no difficulty at all in correctly setting the ignition timing, for example; but this does not mean that the correct functioning of the whole ignition system, for which expensive analytical equipment is needed, can be carried out as well.

For the same reason of lack of suitable equipment, many other jobs cannot be undertaken at home. In this category are the checking and adjustment of the generator charge rate, balancing of wheels and aligning of front wheels, skimming of brake drums, and setting the mixture of emission control carburettors or servicing fuel injection systems. But there are many cases where thoughtful study of the explanations can save cash by enabling the motorist to work out for himself what has gone wrong, and helping in the decision as to whether it is something he can repair himself by buying and fitting a new component.

The 'leave well alone' policy

Some people think it is better not to disturb something that is running satisfactorily, and advise the policy of leave well alone. Perhaps this springs partly from breakdowns resulting from badly-executed maintenance. The man who decides to adjust the tappets of his engine, although they seem to be perfectly all right, and then makes them much noisier than before, may well wish he had not disturbed them in the first place.

The answer, however, is that distinction has to be drawn between leaving well alone, and ignoring an increasing need to attend to something which is becoming daily less well! The secret lies in reading-up and thoroughly understanding what has to be done to keep mechanical components of a car in top condition. In this way cash is saved by reducing the wear rate, extending the life, and forestalling the need for repairs. As additional reward, the man who does the work himself is better equipped to cope with any roadside breakdowns because he knows how things *should* work and can find out what has gone wrong. He can also speak more knowledgeably to garage staff, defining exactly what he wants to be done.

When and what to do

Use of the book in conjunction with the car handbook or, better still, a more detailed workshop manual or owners' service guide for the model, is recommended. Because cars differ so widely in their design and maintenance requirements, I have concentrated on explaining how things work, the different systems likely to be employed in the design of the car, and what is generally necessary for their upkeep.

For the question of when or how often is this or that service procedure to be carried out, the handbook, lubrication chart, or manufacturer's service schedule is invaluable. Because these are often designed to group a large number of service procedures together to be carried out at given mileage intervals by a well-equipped garage, there may often be more work than the ordinary chap can reasonably undertake especially if time available is restricted to a weekend. So, at the end of the book will be found a chronological service schedule, aimed to spread the work of car maintenance over a year; cross-reference page numbers show where the explanation of a given service procedure will be found in the book.

This schedule is a minimal one, and if the maker's handbook recommends that a given task be carried out more often, their advice should be followed and the additional work be fitted in at appropriate times. The schedule aims to combine related jobs – attending to brakes at the same time as wheels are being changed round, for example; and it is seasonal as well, with the idea of doing all work on the car at the right time of the year – for your comfort as well as the car's.

9

Systematic work

It is recommended that the whole procedure for a given maintenance task be studied fully first, so that you can see before starting what has to be done, what equipment will be needed, and what parts may be required. Many people do not have a second car or a spares department within walking distance; so it is logical to obtain what you may need in advance, especially when dealing with components such as brake shoes which cannot be inspected until quite a lot of work has been done. If new shoes have been purchased but prove not to be needed yet, they can be put aside until they are; and any price rise on the component before it is fitted will have been avoided. In the same way, it is sensible to carry a replacement fan belt, set of points and sparking plugs as useful spares which will eventually be needed anyway.

When dealing with unfamiliar components it is a wise policy to note the layout or arrangement before dismantling, the easiest way being a rough sketch, with such notes as: blue and yellow wires to here or spring passes behind this and clips on from rear. The mere fact of writing it helps one to remember, and the notes enable one to check afterwards that everything has been re-assembled correctly as it was before.

On completing all jobs make a final check: everything re-fastened correctly and checked for tightness? No lids left undone? No bricks still in front of wheels? No tools left lying about on wing valances, under bonnet, or elsewhere?

Logical, systematic work, with full understanding of what has been done and why, without any guesswork or leaving to chance, makes car maintenance a satisfying task, and is specially important where the safety of the vehicle on the road is concerned.

2 · In the Interests of Safety

MUCH of the work involved in car maintenance is done in the interests of efficiency – to ensure that the car is reliable and to avoid excessive wear; but there is another, even more important reason – that of safety. In this chapter we deal with all parts of a car which are specifically concerned with its safety on the road. These aspects of car maintenance call for special care, and it is essential that you understand exactly what you are doing, and that faulty workmanship is avoided. It is not intended to frighten one off attempting such jobs as the overhaul of brakes, yet it would be irresponsible to suggest that there is any room for experimentation or learning by one's mistakes. Explanations should be followed carefully and no short cuts taken. Remember that an important part of vehicle servicing is simply that of inspection – a visual check that all is well. If you are undertaking the routine attentions laid down by the manufacturer, and thus avoiding the costs of taking the car to a garage, the onus of regular inspection also falls on you. The annual compulsory vehicle test is much too infrequent to be depended upon for ensuring that a car is still in fit condition.

Wheels and tyres

Tyre wear tends to occur unevenly, so to equalize wear and avoid having to scrap tyres needlessly early because of thin patches on the tread, regular change-round is necessary. Tyres which have undergone a spell at one end of the car wear differently when moved to the other end or the other side. While the wheels are off, one at a time, there is an opportunity to check brakes, suspension components and so on; and so the regular removal of wheels can be regarded as the first essential of car maintenance. Even in such a simple matter as changing a wheel, damage can be caused if you do it incorrectly.

First, correct use of the car jack must be understood. Simple principles of gearing and leverage are used in jacks, to multiply the effort of the man operating it. The sort of screw scissors jack provided with many popular cars may take about fifty turns of its winding handle to give a total lift of some nine inches. The travel of the handle in circular movement will be much greater than may be imagined, and the total multiplication of effort in the order of 100 to 1. This means that a load of 30 pounds applied by the man turning the handle will be increased to a theoretical lifting effort of up to 3000 pounds. The rotary movement of his hand on the jack handle will have been some 900 inches for a mere 9 inches of vertical movement.

Jacks of this kind are arranged to give progressive reduction of gearing towards the end of the travel. At first, it may take only about five turns of the handle to give an inch of lift, but the last inch may take as many as fifteen turns. This is the same as going into a lower gear when driving the car up a hill, and the extra number of turns give more effort for the final stages of lift as the jack takes the entire weight of one corner of the car off the ground. Pillar jacks and hydraulic ones tend to have consistent gearing, which means rather tedious work in the early stages of lifting, when the handle can be moved rapidly round or up and down without giving much vertical movement.

If possible, jacking-up should always be done with the car on a level concrete surface. Both wheels at the opposite end of the car from that being raised must be chocked by a brick or other susbstantial block on the downhill side, if jacking on a slope is unavoidable. When jacking on the level, chocks should be placed on each side of the wheel diagonally opposite to the one which is being lifted, but in this case it is less important to secure both wheels. As will be explained later, it may become necessary to release the handbrake while the car is jacked up, so it can be seen that efficient use of wheel chocks is important, and it is not sufficient just to rely on the handbrake even when jacking the front wheels. If no concrete surface is available, a flat slab such as a paving stone should be placed beneath the jack; otherwise it will sink into a gravel drive, and even into tarmac. Bricks, because of their irregular shape and the possibility of the jack toppling off them, should not be used for this purpose.

Remembering the tremendous effort which the jack will exert – sufficient to lift a corner of the car, although applied over a small area – it is vital to ensure that it has properly engaged with the reinforced mounting point built in by the car manufacturer. The handbook will always explain exactly where and how the jack should be used. Some cars have a tubular mounting point, protected by a rubber grommet. The grommet must be pulled out first, any accumulated mud cleared away, and when the lifting rod of the jack is inserted it must be tapped lightly to make quite sure it has fully engaged.

Before using a screw jack, a few drops of oil should be scattered along the screw threads. They will make the work of lifting easier, and use of the jack will spread the oil along the thread, preventing rust formation before it is next used. Pivot points should also be lightly oiled.

Begin turning the jack until the body of the car is seen to start lifting, and then before proceeding further it makes life easier to slacken the wheel nuts while enough of the car's weight is still carried by the wheel to prevent it from turning.

13

The modern trend in car wheels is to have exposed wheel nuts, but the majority still conceal the nuts behind a chromed hub plate. Care is needed when removing this, especially if there is a surrounding wheel trim as well, since it is all too easy to cause unsightly damage. A special tool for removing the hub plate is provided with the basic tool kit of many current models, and this should always be used; otherwise seek some similar piece of thin, sharp steel, such as a broad knife with blunt end. Avoid the temptation to use a screwdriver, which will inevitably score the paint, damage the hub plate, and cause unsightly blemishes – all too often done by careless mechanics at a service station. The plate is pulled off by a sharp levering movement, with the free hand held against the plate so that it does not go clattering to the ground.

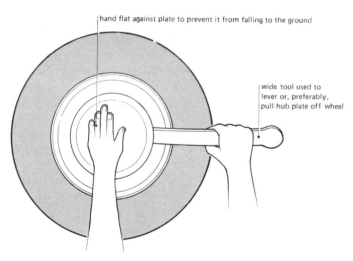

hand flat against plate to prevent it from falling to the ground

wide tool used to lever or, preferably, pull hub plate off wheel

Each of the four or five wheel nuts should be slackened-off one full turn, using the proper wheel-nut spanner provided for the car. There is no possibility of using an open-ended

14

spanner on wheel nuts, so if the wheel wrench has been lost, another must be purchased from any accessory shop. Garage mechanics tend to use a big wheelbrace and tighten wheel nuts unnecessarily tightly; if one of these was last used to tighten the nuts there may be difficulty in undoing them with the standard tool. Avoid hammering; usually the nuts will yield to firm foot pressure on the wheel-nut spanner.

Check that the jack is still correctly engaged, and is clear of the bodywork, and continue jacking until the wheel is about an inch clear of the ground. The wheel nuts may now be unscrewed the rest of the way and removed, leaving the upper-most one in position until last, and placed on a piece of cloth or other clean surface. The popular practice of turning the hub plate chrome-side down and putting the nuts into it causes unnecessary scratching of this piece of decorative trim, and is not recommended.

The wheel can now be removed – try to take the weight and lift it off, rather than dragging it over the threaded studs. Removal of wheels now gives access to the brakes and to the suspension components, dealt with later in this section. It also gives opportunity for inspection of the wheels themselves, and the tyres. Look for damage from kerbs, on the edge of the wheel rim. A small bend does not matter, but severe bending and possibly a split along the edge of the flange calls for special caution. The tyre will have to be removed, and the wheel repaired professionally, or even replaced.

Attention to tyres

Roadside breakdowns due to tyre trouble can often be fore-stalled by careful attention while the wheels are off. Look first at the side walls of the tyre, especially on the inner surface, concealed from view when the wheel is fitted to the car. Slight traces of cracking in the upper surface of the rubber are permissible, provided they are checked at regular intervals against further deterioration. Bulges in the side wall tell of more serious damage, and are not acceptable. A tyre with sidewall damage of this sort is scrap, and must be replaced. If the fault resulted from a flaw in manufacture, an allowance against the remaining tread depth may be claimed from the maker; but no credit will be given if it is shown that careless driving over kerbs has caused the damage.

Now take a pointed instrument such as a screwdriver or meat skewer, and prise out any flints trapped in the tread; some of these can work their way through, if not removed, eventually to cause a puncture. If a substantial nail is seen in a tubeless tyre, the best policy may be to leave it there for the time being. If it has penetrated the tyre, the rubber may well have sealed itself round it, and air will not be lost until it is removed. The wheel may therefore serve as the spare until a proper repair can be carried out. If the nail is taken out beforehand, the tyre will deflate at once and you are left without a usable spare.

On either side of the rim will probably be seen balance weights – a strip of curved lead shaped to sit in the curve of the wheel rim, and varying in length from as little as a $\frac{1}{2}$-inch to 3 inches – located by a small flange folded over the edge of the rim. Check that these are secure. If a loose balance weight is found, its position on the rim should be marked, preferably with a chalk line at each end, on the wall of the tyre, or alternatively with pencil on the rim itself. It may be prised off, the flange gently squeezed towards the lead with pliers, and then re-positioned on the wheel with two or three light blows from a hammer. These balance weights make up for the inevitable unevenness of rubber weight in the tyre, and there are few cars which can manage without wheel balancing. However, this is a professional job, requiring a special machine which will revolve the wheel at high speed and show where the weight should be added, and how much is needed. Whenever new tyres are to be fitted, or a puncture to be mended, re-balancing should also be requested. Wheels which are out of balance cause steering shake and vibration, sometimes leading to visible shake of the facia and scuttle. The critical speed at which the effect of unbalanced wheels is most apparent is usually between 55 and 75 mph. Occasionally there is low-speed vibration, perhaps at 35 mph, experienced again at higher speed, perhaps 80 mph.

The fact that there are no balance weights visible on a wheel does not necessarily mean that it has not been balanced; it may have been checked and found within acceptable limits without addition of any weights. Similarly, the presence of balance weights is not an infallible indication of a balanced wheel. Irregular wear can make it go out of balance; and it has even been known for a new tyre to be fitted to a wheel without removing the old balance weights. The experience of vibration at moderate speeds is a sure enough indication that the car should be taken to a tyre-fitting specialist or a compe-tent garage, with a request for all wheels including the spare to be balanced.

The depth of tread remaining can, and should, be checked visually at frequent intervals; but when the wheel is off there is a good opportunity for a careful check. A proper tyre tread depth gauge calibrated in millimetres should be used; they are available at well under £1 from accessory shops. If no such gauge is available, a rough check can be made with a propelling pencil: extend the lead into a recess of the tread until the end of the pencil lines up with the surrounding rubber, and measure against a ruler. The legally-defined minimum depth is 1 mm, but it is generally agreed that wet-weather safety is considerably reduced once the depth is below 2 mm. The regulations apply to the whole tread width and circumference, so that a bald patch or crown of the tread

17

below 1 mm – or on the shoulder – can render a tyre legally unusable. But it is not unreasonable to regard a tyre as serviceable for the spare wheel when it is in the 1 to 2 mm category. Depth readings should be taken at any apparently thin parts of the tread.

Tyre expenses can be reduced if you check the wear pattern and act accordingly. Tyres often wear on the outer part of the tread more rapidly than the inner part; yet when you change the wheels round the tyre still remains in the same relative position, with the outer part on the outside. Before this wear gets too severe, therefore, the tyres should be changed round on the wheels. If a puncture has to be repaired, it is a good idea to ask for this to be done at the same time, if needed. Otherwise a visit to a tyre-fitting station is worth while, with the request that the tyres be changed round on their wheels. The exception is the Michelin XAS, which has an asymmetric tread pattern. The outside wall of this tyre is marked in four languages, and it must remain outside.

Tyre fitting should be done only by someone who knows what he is doing. The amateur can easily get the inner tube trapped between tyre and rim, while tubeless tyres need to be in a tyre-fitting cage for safety when using a tyre line to spring the beading into position on the rim. As already explained, re-balancing should follow any change of position of the tyre on the rim; so here is another reason to discourage do-it-yourself tyre fitting.

Tyre laws also prohibit the mixing of radial and cross-ply tyres at either end of the car. The only way in which radial and cross-ply tyres may be used at the same time on a car is for two cross-ply tyres to be fitted at front, and two radials at the rear. This is not affected by whether the car has front- or rear-wheel drive, and anyone who has experienced the dangerous instability of a car with these two tyre types mixed the other way round will appreciate the sensibility of this regulation. It must be appreciated, therefore, that identity of these two types of tyre construction is important; if you examine the markings moulded into the side wall, you will find as well as the maker's name, the size of the tyre, the word tubeless if there is no inner tube, and the word radial if the tyre is of radial-ply construction. Radial-ply tyres have a stronger tread base, are less prone to distortion and scrubbing on the road; as a result, they give better road-holding and last at least twice as long as the equivalent cross-ply tyres. Because the side wall is softer, they also tend to give less rolling resistance, hence a small saving in fuel consumption is an additional benefit. So although dearer, they are an economy in the long term.

Ideally, you will have five radial-ply tyres on your car – four on the road and one in the boot as spare. A spare is legally unusable if it is a cross-ply, while the tyres in use are radials. If cross-ply tyres are fitted, the best advice can be to wait until replacement is due and then go over to a complete set of radials. Most tyre-fitting stations will give an allowance on unused tread remaining on any of the cross-ply tyres which have to be discarded at the same time.

Having established that all tyres are of the same type, i.e., all radials or all cross-plies, we can now think of re-fitting in changed order to equalize wear. This can be done in one of two ways, according to whether it is intended to put the spare into service, or whether the spare is a worn but still legal tyre, to continue as such. Whichever is the case, we begin by fitting the spare in place of whichever wheel has been removed, and again it does not matter too much at which corner of the car the rotation process starts. The wheel taken off now goes to the diagonally opposite position. The wheel from here goes to the opposite end, on the same side; and this one goes to the diagonally opposite position. The wheel now removed becomes the new spare; or, if a worn tyre is to be retained as spare, a further change is made at the corner where the rotation process started.

When re-fitting wheels, make sure the nuts are within reach; offer-up the wheel, push it well on to the studs, and fasten the nut to the uppermost stud, finger tight. The bottom of the wheel is then held while a lower nut is put on. The wheel can now be released, and remaining nuts fastened finger tight. The spanner is then used to tighten-up working round them from one to the opposite one, progressively. If a rear wheel is being fastened, it will help to apply the hand brake. Nuts are turned as tightly as is easily possible with the car

Wheel nut tightness should be taken up progressively *on opposite wheel nuts first, then adjacent ones. Final tightening by foot is a good idea, after the weight of the car has been lowered enough to begin to flex the sidewall.*

still on the jack; then the jack is lowered until the tyre wall begins to cushion outwards, showing that the weight of the car is beginning to bear on to the tyre. Now tighten-up hard, again working oppositely in random fashion from nut to nut to ensure equal tension. Finally, completely lower the jack and remove it and do a final check that all nuts are screwed on as tightly as the spanner allows. Now replace the hub plate, using a smart blow with the flat of the hand to push it over the final lug. Never replace the hub plate until you are quite sure that the wheel nuts have been properly tightened.

Tyre pressures

A few car manufacturers recommend the same tyre pressures for all wheels, especially on cars which have fairly even weight distribution between front and rear wheels. These, however, are in the minority, and it is more usual for slightly lower pressures to be recommended for the front wheels. A car such as the Triumph 1500TC has a recommendation of 22 psi (pounds per square inch) in the front wheels, and 26 psi for the rear wheels. The engine of this car is at the front, so not surprisingly the weight on the front wheels is greater than at the rear; and it may seem illogical that the wheels carrying the greatest weight should have softer tyres than the less-laden wheels. The explanation is that the stability and handling characteristics of the car are a more important consideration, and these are benefited by having the front tyres slightly softer than the rear ones.

Correct setting of tyre pressures is important for safety. In a bad motorway accident which I investigated it was apparent that the reason why an Austin 1100 estate car had gone out of control, veered across the centre strip and collided head-on with a car on the opposite carriageway with fatal results for occupants in both of them, was that the tyre pressures had been

21

neglected although the Austin was very heavily laden. As well as the serious effect on steering response which soft tyres can have, there is also the grave risk of tyre failure. As it rolls along the road, a tyre flexes with each revolution, and this produces a certain amount of heat in the carcase of the tyre. Normally this is carried away by the passage of air over it, but the softer the tyre in relation to the weight carried, the more flexing to be done and the greater ¬he heat build-up. Eventually this can reach the point where damage to the tyre is done, and a burst – the dreaded blow-out – can result.

Radial-ply tyres, with their thinner and more pliable side walls, flex more easily than cross-ply ones. They take less effort to push them along the road, and they generate less heat. Higher tyre pressures are usually recommended for radials, to make up for the greater flexibility, and it is important to check that the tyre pressure listed in the handbook refers to the type fitted.

If an error is unavoidable it is better to err on the side of making the tyre harder (higher pressure), rather than softer. Air escapes through the rubber, albeit at a very low rate, all the time, and a pressure drop of about 1 psi per week is quite normal. Better, therefore, to have the tyre a little harder than standard, so that this loss brings it steadily nearer to the correct value rather than away from it. Two further considerations support this view: the first is that garage tyre lines are invariably inaccurate and generally over-read, resulting in an under-inflated tyre, rather than the other way round. The second is that in choosing the tyre pressure figure to recommend for their cars, manufacturers have to consider the comfort aspect. The figure adopted will inevitably be a compromise between the best setting for good steering and handling, and that for best ride comfort. Harder tyres will only mean a slight penalty of suspension harshness on poor roads, and safety will be unimpaired.

22

Because of the likelihood of unreliability in a gauge incorporated in a tyre-pressure line, a pocket tyre-pressure gauge is a good purchase. These are usually remarkably accurate and better to be trusted than a tyre-line gauge, but there is another reason for their use. On the road, tyres warm-up as the car goes along, and a small increase of pressures takes place – the more significant the warmer the weather. Therefore, to stop at a garage and check tyre pressures with the tyres warm can result in the real setting when they are cool being too low. Pressures should therefore be checked with the pocket gauge while the tyres are cool, before driving off in the morning, and the opportunity to add any extra pounds needed should be taken next time the tank is filled. As an example: recommended pressure 24 psi; gauge reading of cold tyre, 21 psi; reading with tyres warm, at motorway service station, 26 psi; therefore increase pressure of the still-hot tyre to 29 psi.

This is very different from the all-too-common and potentially dangerous practice of letting hot tyres down to the recommended figure. The alternative technique, which I adopted while garage air lines were switched off during the 1974 power crisis, and found less effort than I had expected, is the routine of using gauge and foot pump at home to set tyre pressures correctly at least once a fortnight.

It is my personal view that manufacturers are quite wrong to recommend, as they sometimes do, that the same pressures may be used for all conditions; in other words, that no increase is needed for the extra weight of family, holiday luggage, roof rack, perhaps even a caravan on tow. In such conditions, especially if high speeds are anticipated as well, an increase of up to 6 psi all round seems more sensible. Always retain the recommended differential between front and rear tyres, and remember that the pressure of any one tyre must always be the same as that of its partner on the other side, at the same end. The spare should always be maintained about 4 psi

23

above the hardest setting of the tyres on the road, so that it can be let down to the correct value if it has to be fitted by the roadside.

When travelling abroad it will be found that most Continental garages have a tyre line whose gauge is calibrated in psi, as well as in their own metric measurement of kilograms per square centimetre. Occasionally, however, one may come across gauges reading only in metric, for which the following conversion table will prove useful:

Kg/sq. cm	Psi	Kg/sq. cm	Psi
0·95	14	1·83	26
1·12	16	1·90	27
1·27	18	1·97	28
1·41	20	2·11	30
1·55	22	2·25	32
1·62	23	2·39	34
1·69	24	2·53	36
1·76	25	2·67	38

An obvious but easily-forgotten point is that after wheels have been changed round, tyre pressures need to be re-set.

Brakes

Considering their exposure to the weather, the irregular but frequent nature of their use, and the periodic build-up of considerable heat, it is remarkable that the braking systems of vehicles give the reliable, trouble-free service generally experienced. Luckily is it so, since there is no more vital aspect of car safety; few car accidents ever happened which would not have been avoided if one or other vehicle had managed to stop in time. Luckily, too, brakes are, in the main, straightforward, and their maintenance needs are easily attended to.

We have two main types to consider, and two different

operating systems. First we have the drum brake, which has developed over the decades and still has an important part to play as the standard system for at least the rear wheels of the majority of the world's cars. The second is, of course, the disc brake, originally evolved for aircraft use, adapted for high performance cars by Jensen, Jaguar and Triumph, and now standard on the front wheels of everyday cars. A few cars, like the basic version of the Morris Marina, still have drum brakes front and rear; and there are many fast, expensive cars such as Jaguar and Rover which have disc brakes for all wheels. On the majority, however, you will find disc brakes at the front wheels and drum brakes at the rear. This combination has been found to give the best compromise between the need for braking efficiency and the requirement to moderate production costs where reasonably possible. It is, after all, at the front wheels that the disc brake's ability to get rid of the heat developed much more rapidly, and to keep working efficiently, even though very hot indeed, is most valuable.

In the disc brake, the similarity is to a large flat steel plate whirling round between two relatively small pads of friction material. When the brake is applied, the pads move towards the shiny surface of the steel plate or disc, squeezing it between them and arresting its movement. The drum brake comparison is to a dish with a steep rim, the friction material being pushed outwards against the inside surface of the rim of the dish (or drum). Governing factors in brake efficiency are the type of material used for the friction lining, the size of the area which is rubbed by the friction material during one complete revolution (often referred to as the swept area), and the force with which the friction material is pushed into contact with the rubbed face of the drum or disc. As you push harder on the brake pedal, you increase this force, and so increase the braking effort achieved.

From these basic descriptions, which will be common

knowledge to many readers, some relevant points about braking efficiency emerge. The drum brake has a much larger area of friction material than the disc type; yet the actual area of surface rubbed by the friction material is much larger in the disc brake than the drum. Less easy to understand is the fact that if curved friction material inside a brake drum is pivoted at one end, and pushed into contact with it while the drum is rotating, the movement of the drum will actually tend to pull the friction material into even harder contact with it. This effect is obtained only when the rubbed surface of the drum is travelling towards the pivot point of the friction material (or brake shoe, to use the correct term). If the drum is rotating in the opposite direction, away from the pivot point, the opposite effect is obtained, the drum now tending to draw the lining material away from it. In the average brake drum used on car rear-wheel brakes, there are two shoes, each pivoted at one end and separated at the other by the hydraulic cylinder which pushes them apart. One of these shoes has its pivot point

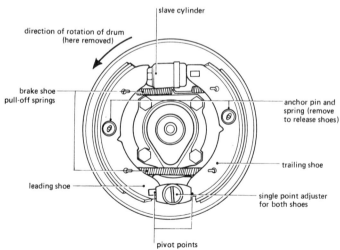

placed so that in normal forward motion the drum surface revolves towards it, with the resulting tendency as explained to increase the force applying the brake lining to it; and this is called the leading shoe. The other one is called the trailing shoe.

When drum brakes are used on the front wheels of a car, the manufacturer often provides two hydraulic cylinders, one for each brake shoe. By pivoting each brake shoe so that the direction of rotation of the drum is from the hydraulic operating cylinder to the pivot, when the car is travelling forward, two leading shoes are provided within the one brake drum. Efficiency is increased, but the drawback is that when the car is reversing both shoes become trailing ones, and the resulting braking effort is greatly reduced. It is for this reason that leading and trailing shoes are retained in rear-wheel brakes, otherwise the car would be difficult to stop when reversing, and the handbrake would be ineffective whenever the car was pointing up a hill.

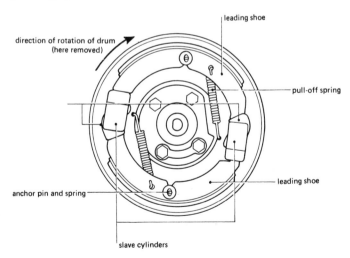

leading shoe

direction of rotation of drum
(here removed)

pull-off spring

leading shoe

anchor pin and spring

slave cylinders

It is important for these differences of brake shoe layout to be understood, because they play a part in brake maintenance. Inevitably, as the leading shoe does more work than the trailing one, its lining material tends to wear out more rapidly. Also, the self-wrapping effect of the leading shoe in a drum brake tends to increase the braking effort achieved for a given push on the pedal. This is called self-servo effect. The disadvantage of drum brakes is that springs have to be provided to pull the shoes away from contact with the rubbed surface of the drum. Each time the brakes are used and then released, the springs pull the shoes back to their pre-set off position; and as the lining material wears more and more, the distance travelled from off to contact increases, and is noticed as reduced efficiency and a proportionate increase in the distance which the brake pedal is depressed before any braking effect is achieved. It is to compensate for this wear that periodic brake adjustment is necessary.

Considerable effort and ingenuity have been expended over the years in trying to make drum brakes which are self-compensating for wear, and need no adjustment. An early attempt at this was seen on the post-war Triumph Roadster, and a more refined system was tried on the Vauxhall Victor 101 series. Many current cars have no requirement for brakes to be adjusted; so it is important to check the handbook first, to see whether brake adjustment has to be done periodically as a maintenance job (look also for Adjust Rear Brakes as an instruction on a voucher for routine maintenance), or whether it takes place automatically.

Brake inspection

Before contemplating adjustment of brakes, we must ensure that sufficient lining material remains; otherwise repeated adjustment would eventually bring the metal part of the shoe itself into contact with the drum, resulting in dangerously ineffective braking, and damage to the drum itself. First jack up the car and remove a wheel; usually brake inspection is combined with the periodic rotation of wheels described earlier. Removal of the wheel reveals the brake unit, which may be of drum or disc type (disc brakes are dealt with later). Expect to find the brake drum itself lightly covered with rust on its outside surface – the result of frequent exposure to road spray and heat from braking hard from high speed, which burns off any protective paint. In the face of the drum will usually be seen two or three small set-screws, which must now be undone. Heat and corrosion often bond them in place, and application of penetrating oil, plus gentle persuasion with a 'soft' hammer may be needed to free them.

Once the set-screws have been undone, the brake drum can usually be withdrawn, being careful to pull it off straight so that it does not go crooked in withdrawal and jam against the brake linings. If a hammer has to be used to free the drum, care is necessary; again, a soft hammer should be used, or the blows of an ordinary hammer should be cushioned by a piece of wood. Take care that hammering does not displace the jack. If such persuasion is necessary, it is a wise precaution to place a support, or even lay the wheel beneath a low part of the chassis so that if the worst happened and the car toppled off the jack it would not crash to the ground. Better still is the use of a proper axle stand to support the wheel-less corner of the car. It may also help if you slacken-off the brake adjustment, before removing the drum – see the later explanation of brake adjustment.

On some cars, as a production economy, the brake drum is cast in a single unit with the surround of the wheel hub. The manufacturer's gain is the owner's loss, as inspection of brake lining is made more difficult. To remove the drum, the hub must be withdrawn as well, for which a hub-drawing tool

29

will be needed. These are sold by accessory shops at about £2 to £3. If removal of the wheel reveals no set-screws in the face of the brake drum it is fairly safe to assume that the brake and hub are in one unit, and a hub-puller will be needed. The central nut at the end of the hub must then be removed, but take care that the split pin is first withdrawn if the nut is of what is called the castellated kind. This is, as implied, like a castle turret, and the split pin passes through a hole in the hub and is then bent round the nut, its head and shank passing through the recesses in the castellated nut. The pin is bent straight with long-nosed pliers, and pulled out, a point to bear in mind being that the efficient mechanic never re-uses a split pin; he always fits a new one each time it has been necessary to remove the nut.

Directions for whichever hub-puller has been purchased should be followed, and the brake drum is pulled off, bringing the wheel bearing with it. As it may be difficult to avoid dropping brake dust on to the tapered stub axle, it may be necessary to clean it carefully with cloth and apply fresh hub grease before re-assembling. If the centre nut is not apparent, it may be concealed behind a cap which is prised off with a screwdriver. Bear in mind that removal of this nut is un-necessary if – as is the case on most cars – the drum is separate from the hub.

Removal of the drum reveals the brake linings for inspec-tion. They are usually fastened to the shoe by means of rivets, and the critical stage for replacement is when there is any risk of the metal rivet coming into contact with the inner surface of the drum. The wear rate is fairly slow in most cars driven reasonably, and so clearance of 1 to 2 mm at whichever rivet is nearest the surface is acceptable for further service. If the rivet nearest the surface of the lining is closer than this, and especially if brightness in the face of the rivet suggests that it is already beginning to make occasional contact with the drum, renewal of linings is necessary. On some popular makes a bonded brake-lining is used. Judgment as to whether replacement is necessary is less easy, but look for any suggestion of separation of the lining from the shoe, any traces of cracking of the lining, or for thickness reduced to a minimum of about 2 to 3 mm. In view of the importance of good brakes, the best advice is to replace brake linings if there is any doubt about their ability to last to the next service attention.

Presuming that we have decided the linings to be fit for a further spell of use, we move on to re-assembly; but first the inside of the brake drum should be cleared of all brake dust by wiping with clean cloth. While the system is revealed to view, examine also the flexible brake pipe for any signs of cracking, and the slave cylinder for any suspicion of fluid leakage. These matters are covered in the section on Brake Hydraulics. Following removal of brake dust from drum and brake assembly, the drum is replaced, noting that only in the correct position will the holes in the drum line-up with the threaded holes in the hub flange on some cars.

Individual instructions in the handbook for the car should be studied to locate the position of the brake adjustment nut. Generally, however, this is a square-shank nut protruding from the lower part of the brake back-plate; and a spanner should be located which is the correct size to fit it. Make sure that the handbrake is fully released, and turn the adjusting nut. This revolves an eccentric cam inside the brake, pushing the shoes farther out, into contact with the drum. On most cars, clockwise rotation of the adjuster screw tightens the brake, and it should be turned, whilst checking intermittently for rotation of the drum, until it is locked and no further rotation is possible. The adjuster moves in clicks or notches, and it should now be backed-off (rotated anti-clockwise). The drum should now be free to move, but if not, the brake should be slackened until it is no longer rubbing.

Do not confuse brake friction with the normal quite considerable drag of a back axle – always reluctant to turn by hand through the differential while one wheel is still on the ground. Slight hiss of the brake rubbing lightly is not a cause for concern – it is probably just a small high spot which will quickly wear off in service. As a general guide, if the drum can be turned by hand without too much effort, the brake is unlikely to be too close-adjusted; though it could be too loose. Re-check by tightening on one more notch, and slackening back again if necessary, before re-fitting the wheel.

The explanation given earlier about two-leading-shoe brakes now comes into the matter of adjustment. A two-leading-shoe system will be identified by the presence of two hydraulic cylinders, revealed when the drum is removed; and in addition to the flexible hydraulic pipe at the back, there will be a small rigid hydraulic pipe taking the fluid from the region of the flexible pipe to the other end of the brake assembly. The important point is that there will also be *two* brake adjusters, each to be dealt with individually. As already explained, twin slave cylinders and two leading shoes are most likely to be found when a car has drum brakes at the front.

When satisfied that the brake is in optimum adjustment, the procedure is repeated at the wheel on the other side of the car. Never adjust one side only.

Re-lining brakes

Most parts stores sell brake shoes fitted with new linings, and credit is given later when the old ones are handed in. Done correctly, the fitting of new brake linings is an efficient and easily-managed task. First, slacken off brake adjustment to the limit; now proceed as in Brake Inspection to the removal of the drum, and identify the small cylindrical unit positioned between the ends of the brake shoes. This is the slave cylinder, which operates the brakes; to ensure that its piston does not fall out while the shoes are being removed, it should be tied longitudinally with a piece of string. The arrangement of springs, and the holes in which they are fitted should now be noted in a rough sketch. Sometimes, similar yet functionally different springs are colour coded; so colours should be noted as well.

If the brake being dealt with is of two-leading-shoe design, there will be two hydraulic cylinders. In practice, only the one whose piston faces downwards needs to be tied, but it may be considered wise to secure both of them.

Removal of brake shoes, here on an Opel Kadett, can be done by gently levering, or pulling, the shoes away from the springs. If the slave cylinder is arranged vertically, instead of horizontally as here, it is a wise precaution to tie it temporarily with string to prevent the piston from dropping out while the shoes are removed.

33

Look for anchor pins with a small coil spring and cup, in the side of each shoe. This is removed, usually by simply depressing and turning the spring cup so that the flat end of the pin lines up with the slot in the cup. The springs and cups are withdrawn and pins pushed clear. Brake shoes can now be removed. Instead of attempting to prise the ends of the springs off the shoes, it is easier to pull the upper brake shoe bodily away from the other, overcoming the tension of the pull-off springs until first one, then the other end of the shoe clears the abutments.

While the shoes are off, the back plate should be cleaned thoroughly to get rid of brake dust. The adjuster cam and the ends of the new brake shoes should be lightly smeared with the special white grease available from accessory shops for brake components. Do not use ordinary grease for this purpose as it may melt when exposed to the heat of severe braking and get on to the brake friction faces.

Generally, both brake shoes will be renewed at the same time; but it has been explained that in a leading-and-trailing-shoe brake, the leading shoe wears more rapidly than the trailing one. Provided it is checked carefully that the shoes are identical in construction, as is usually the case, it is permissible to replace a worn leading shoe and retain the serviceable trailing shoe. Whichever procedure is chosen, the same must be applied each side; never fit one or two new linings on one side only.

To re-assemble, fit the lower shoe into position against its abutments, making sure that – if only one leading shoe is being renewed – the correct one is offered to the lower position, and secure it with the anchor pin and coiled spring. Now position the brake pull-off springs carefully in the correct order and in the correct holes, with the second shoe held outside the brake, overlapping the abutments. It is now fairly easy to pull it against the spring tension far enough for the ends of the shoe to be slotted into the abutments. Check for correct arrangement, location and fastening of the pull-off springs, and now secure the second shoe with the anchor pin, spring and cup. Make final confirmation against the manufacturer's drawing, or your rough sketch, that spring hooks have properly engaged, are in the correct holes, and that the ends of the brake shoes are properly engaged in the abutments provided at the hydraulic cylinder and the cam adjuster.

String may now be cut away from the cylinders, and the brake drum carefully replaced; the position of the shoes may have to be altered slightly to take the drum. Brake adjustment is now carried out, followed by shoe exchange on the other side.

Handbrake mechanism

It will usually be found that adjustment of the rear brakes will automatically improve handbrake efficiency, since in most cases the same brake is used for the parking brake; it is only the means of operation which is different, and to comply with the Construction and Use Regulations this has to be independent of the main brake. This is the second operating system, referred to earlier. Some of the more expensive high-performance cars have an independent brake for the handbrake, as well as independent means of operation, and the big Citroens have a foot-operated handbrake (better referred to as the emergency, or parking, brake), which operates on the front wheels.

The most usual arrangement is for cables to be used in conjunction with an open wire or rod for the front part of the mechanism. Cars such as Vauxhalls, which have a 'live' axle at the back, may have rods mounted actually on the axle, but where there is an independent rear suspension, as on the Triumph Herald, enclosed cables are used. In service, wear can

35

develop in fulcrum connections, where the handbrake mechanism changes direction, and cables can stretch. So adjustment becomes necessary from time to time, revealed by excessive free travel at the lever even when the rear brakes have been newly adjusted.

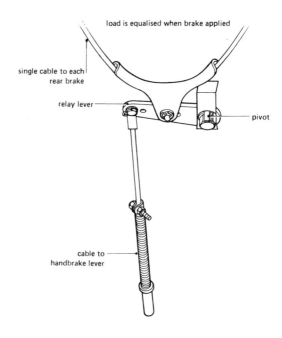

load is equalised when brake applied

single cable to each rear brake

relay lever

pivot

cable to handbrake lever

Examine the complete route of the mechanism from its connection with the operating lever to the rear brake units. Adjustment is usually provided by means of threaded rod and twin nuts at the end of the cables (if any), *and* at the connection with the handbrake lever. To adjust, slacken the lock nuts and turn the threaded part of the operating rod; often the rod has a spanner recess to enable it to be turned, and lightly oiling the threads will help. Adjustment should be made so that slack is taken up, and the first notch of lever movement should begin to operate the brake. Check after adjustment that the rear brakes are still free.

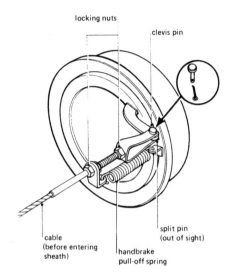

The handbook or lubrication chart should be studied to see what parts of the handbrake mechanism require periodic lubrication. The harness of sliding cables may be provided with grease points, and swivelling pivots for the mechanical linkage need greasing if a grease point is provided, or alternatively a few drops of engine oil should be applied. Grease should be applied to the handbrake ratchet.

Some cars, such as the Morris Minor, have simplified handbrake adjustment by having adjuster nuts within the body of the car, accessible near the handbrake lever fulcrum; and the handbook should be consulted to see whether such straightforward adjustment provision is made on your car. The need to check against inadvertently over-tightening, causing the rear brakes to bind, is again important.

37

Disc brakes

In most respects of service, disc brakes are easier to deal with than drum brakes, and they have the advantage that the state of wear of the linings (called pads) can be examined once the wheel has been removed, without need to dismantle. The brake assembly is called the caliper, and it comprises a casting containing twin hydraulic cylinders, whose pistons bear against the back plates of the pads. In most disc brakes the back plates are flat, and there is no self-servo effect which, as we have seen, is present in the leading shoe of a drum brake. For this reason there is no need for pull-off springs. The slight irregularity of the disc, or lateral movement of a few thousandths of an inch allowed by the wheel bearing, is sufficient to kick the pads out of the way after each application; and they remain in readiness, practically in touch contact with the face of the disc.

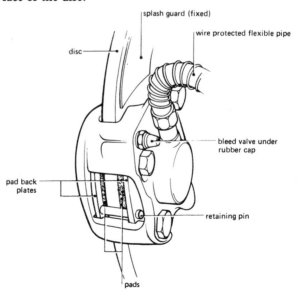

38

They will also go on wearing until all the friction material has worn away, though the metal back plate would then contact the smooth face of the disc and damage it. Before this stage was reached, the thinness of the lining material would allow excessive heat to be transferred to the brake fluid, with risk of boiling the fluid and serious loss of braking power. So it is an essential service procedure to inspect the pads and check the thickness of material remaining. As a broad guide, regard 3 mm thickness as time to replace, and 2 mm as getting dangerously thin.

An unusual type of disc brake, called 'swinging caliper', as fitted to the Fiat 132. The caliper body may have to be removed for renewal of the disc brake pads, though inspection is possible from above. The pads are wedge-shaped, and are renewed as a pair.

39

Access is often rather difficult for measuring, and it is needlessly troublesome to remove the pads for measurement. Instead, measure the gap between the inner faces of the metal back plates. A pair of dividers, obtainable from any good stationer, helps here. Now measure carefully the thickness of the brake disc at any convenient point around its circumference. (A note of the figure, kept in your service record book, will save you the trouble of measuring it again on any future occasion.) Now subtract the thickness of the disc from the gap between the back plates, and divide by two. The result is

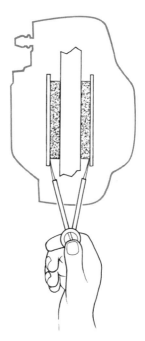

geometry dividers, or a steel rule, can be used to measure total gap between disc brake pad back plates.
Thickness of the disc is deducted, and the result divided by two, to give average pad thickness.

the *average* thickness of the two pads. As wear can take place unevenly, a visual check and, if necessary, individual measurement using the dividers again, is a wise precaution.

If uneven wear is spotted in the early stages, it is permissible to change the pads around, having checked first that they are identical in construction. The more rapid wear, usually of the inner pad, will then have longer to go on the thicker pad. (Refer to Disc Brake Pad Renewal.)

A few cars now have an electric current in a wire embedded in the pad itself, which completes a circuit and lights a facia warning lamp when the brake pads pass a certain point of wear. Check the handbook to see if this is the case with your car, but even if such a warning device is fitted, it should not be relied on; periodic inspection should still be made, in case the warning system itself has gone wrong. Another technique, used by Opel, is to fit a metal separator which holds the pads apart when a certain degree of wear has been reached. As wear takes place gradually, a progressive deterioration tells the driver that pad renewal is necessary, even if he has not been making periodic inspections.

Disc-brake pad renewal

Inspection of disc-brake pads can be made every time a front wheel is removed, without any need for further dismantling; and as the pads remain in light touch contact with the rubbing faces of the disc, there is no need for any routine adjustment. Renewal of the friction linings (pads) is also easier than with drum brakes. Start by removing the front wheel, and examine the retaining mechanism in what is called the brake caliper – the main assembly which locates the pads.

The method of retaining the pads differs from one brake manufacturer to another, but it is fairly general to employ retainer pins. These are better described as rods, which pass

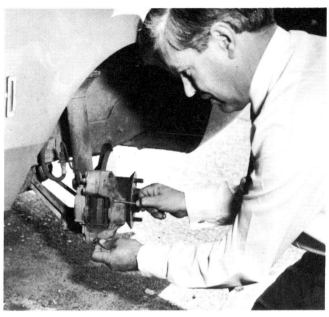

Brake pad location varies according to manufacturer of the disc brakes. On this Ford Capri a split pin must be removed first from the end of the locating rods—then the two rods are withdrawn. Care must be taken that the anti-squeal springs do not jump out.

from one side of the caliper to the other, and either pass through holes in the metal back plates of the brake pads, or across the outer edge of them. There will also be spring clips or small pins to locate the main retaining pins, and these must be removed first and put carefully aside. The retaining pins can now be withdrawn; because of corrosion and heat, a little persuasion with gentle tapping from a rubber mallet may be necessary. The friction pads may now be lifted out. If they are reluctant to budge, they may have formed a corrosion bond

After removal of locking pins, it should be possible to withdraw disc brake pads quite easily; but sometimes they are held by a bond of corrosion and carbon dust. Oil must not be used, as it may attack brake fluid seals. Careful use of a hammer, or levering, is usually successful in freeing them.

with the operating piston, which can usually be freed by levering the pad gently outwards, away from the disc, and then giving it a sharp tap in the opposite direction, back towards the disc. Avoid using force, and do not be tempted to use penetrating oil, which can damage the hydraulic seals.

Before the new pads can be inserted, the slave cylinder pistons must be carefully levered outwards, away from the disc, to make way for the greater thickness of the new pads. An obvious but vital point is to check that the pads are put in the correct way round: black, carbon-like friction material against the shiny surface of the brake disc, metal side away from the disc. Now re-fasten the brake retaining pins and the

43

fastening clips or pins. Proceed to the disc on the other side of the car; brake pad change-round or renewal must always be done on both sides.

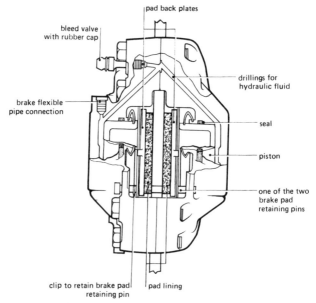

Cutaway view of typical simple disc brake assembly.

Brake hydraulics

After various developments with rods, cables and other systems, hydraulic operation has become universally adopted by all manufacturers for vehicle brakes. Vital points to remember are that the correct brake fluid specification is essential, that no significant drop in fluid level is permissible, and that cleanliness is important. With the aid of the car handbook, locate the brake master cylinder. On some pre-war cars it used to be hidden away under the floor, but it is now almost universally accessible under the bonnet, mounted on the scuttle in the approximate vicinity of the brake pedal fulcrum. A good refinement on modern cars is the use of a transparent reservoir, with a mark showing the recommended level. On others, the screw cap must be cleaned with cloth and then

Modern brake master cylinders usually have a transparent reservoir, enabling the fluid level inside to be seen without need to remove the cap. If this is not the case, care must be taken to wipe the cap first, before removing, to prevent ingress of dirt.

carefully removed to inspect the level. If the level remains about the same for month after month, all is well; but a sudden drop in level is a serious danger sign, telling of leakage somewhere in the system, which must be investigated without delay. Don't just top-up, which is curing the symptoms without dealing with the cause of the trouble.

Brake fluid is not everlasting; nor are the flexible pipes and seals of the slave and master cylinders. Fluid should be renewed annually, and metal brake pipes should be inspected for corrosion or damage. Flexible brake pipes and seals need to be renewed approximately every third year in the life of the car. The work involved in changing the latter is somewhat specialized, requiring knowledge or information about the precise order and arrangement of seals and glands. This will usually be entrusted to an agent for the make, but it is not to be omitted if one wants dependable brakes. The annual renewal of brake fluid, however, is well within the scope of the home mechanic. (Bear in mind that the third renewal of brake fluid will be carried out automatically when seals and flexible pipes are replaced.)

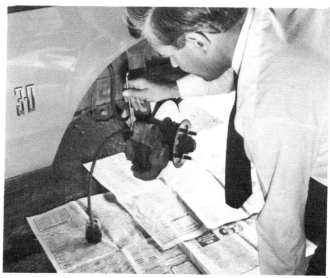

Brake bleeding is an important annual maintenance task. Ideally it is a job for three people—one to press the brake pedal, one to check that no more bubbles of air are being driven into the glass bottle, and one to keep the master cylinder topped up. Final tightening of the bleed valve should be done while the assistant is depressing the brake pedal.

A 2-foot length of small-bore rubber pipe, a jar, and a sealed 1-litre can of brake fluid of correct specification for your car are needed; quote the model's chassis number when ordering, and buy from a reliable stockist, preferably an agent for the make.

It will help greatly to have the assistance of one or even two people – one to operate the brake pedal, another to keep the master cylinder topped-up, and yourself to control operations and work the bleed valve. Usually there is one bleed valve at each brake, resembling a grease nipple, with hexagonal base. It should be cleaned carefully. Then, select a spanner which will fit it, and release by giving about two turns anti-clockwise. The pipe is now fitted over the end of the bleed valve, and fed down into the glass jar. Several strokes of the brake pedal are given, which will force fluid out into the jar, and the level in the master cylinder is maintained by pouring in fresh fluid from the sealed can. When it is estimated that approximately twice the contents of the master cylinder have been driven out into the jar, a final slow depression of the brake pedal is made while at the same time the bleed valve is screwed firmly home. The process is repeated at each wheel, except that after the first bleeding operation, evacuation of approximately one master cylinder's contents is sufficient for subsequent brake cylinders.

Make a final check that all bleed valves are tightly screwed home (though do not over-tighten), and top-up the fluid to the mark in the master cylinder. With the car at rest, check that there is a firm feel to the pedal. If it is spongey and does not come up against positive resistance, further bleeding will be necessary. Look for bubbles as the fluid is forced into the jar, which reveals that air was trapped in the system. Air can find its way in if the valve is not closed while fluid is actually being forced through it, or if the master cylinder is allowed to become empty during the bleeding operation.

Care should be taken to avoid dropping brake fluid on the bodywork, as it is an effective paint stripper. After this operation keep a careful check on the brake-fluid level for the next few days, so that any warning signs of leakage from the bleed valves will be spotted without delay. When the operation has been successfully completed, any fresh fluid left in the can may be retained for topping-up (though beware any need for substantial topping-up); it should then be discarded and not used for the next year's brake-fluid renewal. The old fluid in the jar must never be re-used.

In the interests of safety, modern hydraulic systems tend to be separated, so that if leakage occurs anywhere, partial braking is still possible. The popular way of doing this is with what is called a tandem master cylinder – effectively a reservoir with two compartments, arranged so that fluid can spill over the central dividing wall – and ensuring that real separation only occurs in event of failure. Twin pistons are used, and again in event of failure there will be an increase in pedal travel, but the failed piston will not prevent the other from working. When bleeding the system, make sure that the fluid level is maintained in both compartments. Do not confuse a twin hydraulic brake circuit layout with the other popular arrangement of using twin hydraulic systems – one for the brakes and one for the clutch.

Brake refinements

Because, as explained, there is usually no self-servo-effect with disc brakes, it is often necessary for the manufacturer to provide alternative means of reducing the effort of working the brake pedal, the most popular arrangement being, of course, the vacuum servo. This converts the suction effect of the engine into work, assisting the driver's effort on the brake pedal; the harder he presses, the more servo assistance is provided. In the main, these give excellent service and require no routine maintenance. However, look for a bleed valve on the servo itself; these are sometimes fitted when the servo is combined with the master cylinder, in which case bleeding should begin here.

To allow more braking at the front wheels without fear that the rear wheels will lock-up, causing a skid, some manufacturers fit a valve in the hydraulic circuit to the rear brakes, which reduces or limits the pressure admitted. These also require no maintenance and seldom go wrong; but the presence of a pressure control valve calls for special care to work the pedal slowly when bleeding the rear brakes. As some of them are influenced by the attitude of the car, it may be necessary for bleeding to be done with the wheels fitted and the car not jacked up. Drum-brake bleed valves are invariably found on the inside face of the back plate.

Suspension and steering

Every time the front wheels are removed, the opportunity should be taken to inspect the front suspension so that any developing fault is spotted before it develops into a potentially dangerous failure. Special attention is to be paid to knuckles; these are the joints between steering arms and tie rods or the steering box itself. To allow for vertical movement of the wheels on the suspension, these joints allow a certain amount of rotary movement, so it should not be thought that because they can be turned through a few degrees, there is necessarily something wrong. But no lateral play is permissible; this can only result from wear or damage, which calls for attention by a competent mechanic.

The handbook or the lubrication chart should be studied to see if any lubrication points are provided requiring periodic greasing. At one time, these were the plague of routine service,

49

but beginning with the introduction of the Triumph Herald, they have been progressively reduced and most cars now have no grease points on the suspension, though there may be some on the sliding joints of the propeller shaft (see Drive-Line Lubrication). Some cars have grease points on the swivel pins on which the front wheels are mounted, although there may be none on the steering linkages. Any grease points present should be cleaned carefully, and then the grease gun is used to try to force in more grease. On some well-sealed joints, it will be found that no more grease can be persuaded to go into the joint, and the grease nipple should be unscrewed to make sure that it is not blocked. A sign that all is well is if beads of grease appear around the edges of a joint when the gun is used.

Two steering systems are in general used. By a long way the better of the two is called rack and pinion, which can be likened to the rack and cog-wheel system used for a mountain railway. Within limits, it compensates automatically for wear, and it generally provides more positive steering. Lubrication is usually by means of grease gun (or oil – check the lubrication chart); but to prevent ingress of dirt and grit, the general practice is to fit a blanking plug instead of a grease nipple. Consult the handbook, if this is suspected, and make quite sure that the nut being removed is in fact a grease hole plug. It and the surrounding area must be carefully cleaned; it is then taken out and a grease nipple – purchased from an accessory shop, or borrowed from another grease point on the car – is screwed into the hole, and the grease gun is applied. Then the plug is re-fitted.

The other steering system is the use of a steering box, in which the rotary movement of the steering wheel is translated into to-and-fro movement by means of a worm-and-nut or re-circulating ball system. These usually require lubrication with heavy gear oil, and there is likely to be a filler plug on top of the steering box for topping-up. Again, be quite sure that the nut identified really *is* a filler plug (usually it will be of tapered pattern, with several threads standing proud); clean thoroughly first, then unscrew, and pour in oil of the correct grade until the level is up to the top. Replace the filler plug, being careful not to cross the threads, and to avoid over-tightening. Power-steering systems have a reservoir which must be topped-up with the recommended fluid.

With the front wheels fitted, and the jack removed, an assistant should be asked to waggle the steering wheel to-and-fro while an examination by means of inspection light is carried out underneath. Look for excessive movement of a steering rack on rubber mountings which may be found to have cracked or perished, and for play in any of the steering joints. If anything of the sort is suspected, professional attention should be enlisted.

Another job which calls for a trained mechanic with special equipment is the periodic check on front wheel alignment. With wear, and no doubt the occasional nudge against the kerb, front wheels can go gradually out of alignment, which increases tyre wear. It is a good idea, therefore, for the man who carries out his own car maintenance to ask for wheel alignment to be checked, every time – or at least every second time – the car is submitted for its annual safety test. As wheel mis-alignment does not necessarily affect safety, it is not normally checked as part of the compulsory vehicle test.

Some leaf springs, as used at the rear on the majority of cars, benefit from occasional oil spraying; but on some with rubber separators this is unnecessary and can be positively harmful. Here, too, the handbook for the model should be consulted, to see whether rear suspension lubrication for springs or linkages is necessary or not.

The other aspect of suspensions requiring occasional attention is the shock absorber, referred to collectively by engineers as dampers since this more closely describes their true function. Lever-arm shock absorbers, as fitted on Morris Minors, need an occasional check to be made on the fluid level, topping-up with correct grade if necessary, and again taking care that dirt is not allowed to fall in when the cap is removed. However, these have largely been superseded by telescopic shock absorbers which, as well as being generally more efficient, require no routine maintenance. Out of sight should not mean out of mind, though; shock absorbers are often due for renewal after about 30,000 to 40,000 miles. Their deterioration is indicated by poor ride, with harshness on bumps resulting from wheel bounce, or wallowing and floating on undulations, which can be upsetting for the occupants, especially in the rear of the car. In this event, replacement of the shock absorbers should be requested, and this work is also out of the scope of most amateur mechanics.

Underbody protection

By use of the highest-quality materials and techniques of plating and protecting, Rolls-Royce make their cars pretty well impervious to underbody corrosion; but it is a problem for almost every other car on the road. In the early stages of corrosion, the damage is of a minor nature, affecting the appearance of the car due to rust bubbling through the outer surfaces of body panels such as door edges and wing crowns, from the inside. It is at a later stage that structural decay develops, affecting the safety of the car.

Salts and corrosive substances accelerate this process, and these are particularly active after they have been allowed to accumulate during the winter months. Ideally, the underneath of the car should be hosed off at intervals throughout the winter, especially after spells of ice or snow, when roads will have been salted. However, the aim being to set standards of car maintenance which are reasonably practicable, a realistic target of underbody cleaning in the spring is advised. If this is not done, the mud deposits in wings and crevices will be saline and absorb moisture from the atmosphere, rapidly increasing the rate of rust. A high-pressure, narrow-jet, hose is needed to remove all this surplus mud, and it may even be necessary to use a blunt instrument such as a wooden rod to break away some of the bigger mud deposits.

Removal of mud or accumulated dirt, which otherwise acts like a sponge against the metal, keeping it permanently wet, will help greatly. However, mere exposure to moisture will also allow corrosion to take place every time the underneath of the car gets wet. The best solution is the application of underbody protective coating when the car is new, but presuming that this was not done, a second best can be do-it-yourself application of a proprietary underbody compound.

These are not expensive – a tin containing enough for the average car to be treated liberally may cost only £2 to £3, and the life of the treatment should be at least 3 years. Excess rust and dirt should be removed with a wire brush before applying; but very thorough preparation of the surface is not necessary as it adheres well to lightly-rusted metal. By keeping the moisture at bay it will then inhibit further rusting.

It is essential that the surface be dry, so application during a warm spell in summer is recommended. The compound can be slapped on with an old paint brush, and it can cover metal brake pipes as well, though care should be taken to keep it off flexible brake pipes and union connections. The odd splash falling on the exhaust pipe does not matter, but avoid deposits on the propeller shaft, which can upset its balance. Keep an eye open for any intentional drain holes provided by the manufacturer; if you block them off, water drainage will be obstructed and the corrosion one has tried to avoid will be encouraged.

53

Lamps and electrical equipment

Good night-driving vision, reliable lighting and efficient
windscreen wipers are important contributions to safe winter
driving. Sidelamps are most easily checked with regularity,
by noting their reflection when you stop at night behind a
taxi or other square-backed, dark-coloured, car and tail lamps
can be checked each time the car is parked with lights on. A
faulty stop lamp can be misleading, and a routine should be
adopted for checking these at reasonably frequent intervals.
Headlamps remain in adjustment for a long time, but should be
re-set whenever it is felt that they are shining too low for good
illumination, or dazzling oncoming drivers. On some modern
cars, particularly Continental ones, very easy adjustment
provision is made. In contrast, the trend on British headlamps
for a long time has been to make adjustment awkward with the
idea that they are then less likely to be tampered with. This
is a job best left to a garage, since the setting of lamps is done
by special alignment gauges designed to pick-up against the
three pips moulded into the face of the glass.

If you decide to carry out headlamp adjustment, see first

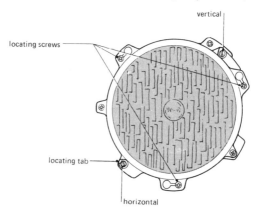

whether the handbook gives advice on how it is done. Many Lucas headlamp units are revealed by undoing a screw at the bottom and removing the surround; two adjusters will then be seen, not to be confused with the three mounting screws. The lock nut on top must be slackened first, and care is needed when attempting to turn the adjuster screw in case it has corroded at the base. Penetrating oil should be used; too much force will easily break the adjuster screw. Once they are free it will be found that the lower screw adjusts horizontally (left-right), and the upper one adjusts vertically (up-down).

A suitable venue for headlamp adjustment is a dead-end lane or a marked-out car park on Saturday or Sunday night; in each case, the ground must be critically level, and a straight line helps. A rug or old coat should be taken along to the chosen venue, to enable the other lamp or lamps to be covered up while each one is being adjusted. Before setting out, the locknuts should be slackened-off slightly, and a check made to ensure that the adjuster screws are free to turn.

If it is noticed during daytime that one of the headlamps appears to have gone black internally, this is a sure sign that one of its sealed filaments has blown. Switch on to check, and if it has failed a replacement bulb or, if a sealed unit, complete new headlamp must be fitted. Re-adjustment will probably be necessary.

Warning of a failure in the indicator lamps circuit (or a flasher bulb) is given by very slow (or sometimes, very rapid) flashing of the indicator's repeater tell-tale. Even the most elementary handbook usually shows how to renew bulbs in the flashers or side and tail lamps.

Dynamo and alternator

Generation of electrical current is an essential function in every car, since without replenishment the battery can go flat in as little as an hour at night, or a couple of days of daylight running. The two common generators are dynamos and alternators, the latter having much greater output, especially at low engine speeds. Both give, in the main, reliable service and require no maintenance, except that some dynamos have a lubrication point for the rear bearing. Check the handbook to see if this is the case, and if so avoid over-oiling. Two or three drops of thin machine oil (or engine oil if recommended) once a month are better than too much oil too seldom. Regulation of the amount of current supplied is governed by the voltage regulator, and this can, in due course, get out of

Alternator drive belts need to be pretty tight. An inch of 'give' under light pressure at the mid-point between pulleys is about the most that is allowable, otherwise slip and shriek may occur, leading to premature drive belt failure.

56

IN THE INTERESTS OF SAFETY

adjustment. If it delivers too much charge, the result will be rapid loss of water from the battery; not enough charge, and the battery will always seem to be weak or flat. Adjustment requires a voltmeter, and is a job to be left to an electrical service agent or competent garage.

Almost without exception, the generator is driven by the engine by means of a 'V'-belt, so called because it is of approximate 'V' section if cut through. In service, it stretches and needs periodic re-tightening, which is described under Driving the Auxiliaries, p. 64. Too little tension causes belt slip and eventually leads to loss of charge; too much can lead to premature failure of the fan belt or the water pump bearing.

Battery

After 2 or 3 years' use, the average car battery has 'had it', and replacement in good time will avoid the misery and unreliability which can result from protracted use of a worn-out battery. In service, its life can be extended by elementary upkeep. The chief point for attention is care over the matter of keeping the electrolyte (mixture of acid and water inside the battery) topped-up. The screw caps are removed, or one-piece cover lifted off, and pure water is poured into each cell as required to keep the level well above the top of the plates – but not so high that it is effectively full. Distilled water for this purpose can be purchased from any chemist – you should take along your own clean bottle. Some people use the melted ice off the ice-box of a refrigerator for the purpose, but it should first be filtered through a fine strainer to remove condensed fat which may have collected out of the atmosphere.

Any deposits of corrosive salts, like white fungus, should be removed from the terminals or tops of the battery, taking care that it does not fall down into the bottom of the battery tray, and proprietary anti-corrosive jelly used to prevent further growth. Vaseline between the two terminals and their connectors will ensure that they do not seize in position, and ensure good electrical contact; petroleum jelly is conductive.

A clean screen

Completing the service attentions needed to maintain a car in safe condition, one should not overlook the periodical cleaning of the inside of all windows, especially the windscreen. It must be accepted that windscreen wiper blades have a limited life, and replacement in the autumn is a good move for safety and clear vision throughout the winter. Use of worn-out windscreen-wiper blades can result in damage to the windscreen, when the metal surround comes into contact with the glass and scratches it.

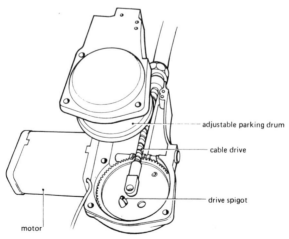

Exploded view of Lucas wiper motor.

Sleepy wipers, failing in mid-travel, are often due to no more than excessive resistance in the drive cable. The motor usually has a drum-shaped cap, whose position in relation to the cover should be marked with a pencil, before undoing the securing screws and removing the lid as it determines the parked position of the wipers. The wiper cable can then be cleaned, re-lubricated with fresh grease, and re-assembled in the correct position. If the wipers are not parking in the correct position, the solution with most wiper motors is to slacken the mounting screws and turn the drum until the proper switch-off position to park the wipers at the end of their stroke is found by experiment; then re-tighten. Other than these adjustments, wiper motors need no regular service; but the spindles on which wiper arms pivot should be given two or three drops of oil at regular intervals.

Clean water in a clean container should be used for topping-up windscreen-washer reservoirs; if it is suspected that dirt may be present it is well worth using cloth – even a handkerchief spread over the opening – as a simple filter. Otherwise dirt quickly blocks the jets, causing endless irritation and potential danger when driving on salted motorways. To clear a blocked jet, use an old toothbrush or a Primus-stove pricker. If the trouble persists it may even be necessary to pull the plastic pipe off the jet entry, flush clean water through the system, and even in some cases to remove the jet itself and blow through using a garage air line. A proprietary additive in the screen-washer bottle plus generous quantities of methylated spirits in winter help for clear vision which can be the key to safety when driving on a wet winter's night.

3 · Under the Bonnet

In a remarkable statement on the standard of maintenance of the average British car engine, Shell claimed that poor tuning and neglected ignition were wasting some 300 million gallons of petrol every year. It seems a sweeping condemnation – perhaps tending towards exaggeration – but there is certainly no lack of cars with inefficient engines on the road. You can identify them from behind, both by the traces of black smoke emitted from the exhaust under acceleration, and from the characteristic smell of unburnt exhaust gases, like that of a car which has just been started-up from cold.

Correct engine maintenance is straightforward and quite satisfying work, with efficiency and reduced running costs as its own reward. Familiarity with ignition and carburation will also equip you to sort out and rectify any engine breakdown on the road, while good maintenance will make anything of the kind less likely.

Lubrication

Because of its importance to the life of the engine, lubrication is the starting-point for good engine maintenance. The oil in the sump, which is circulated under pressure by a gear pump all the time the engine is running, becomes contaminated in service. Tiny particles of metal worn from the moving parts of the engine combine with accumulated carbon and deposits of combustion to make the oil progressively dirtier in service, finally becoming quite black and smelly. The more worn the engine is, the more rapidly the process of contamination takes place due to imperfect sealing of the valves and pistons, allowing combustion products to mix more readily with the oil. Also, the more short journeys and cold starts that are made, the more rapidly the oil becomes contaminated. In cold starts, unburnt petrol washes past the pistons, mixes with the oil and reduces its lubricating effectiveness. In sustained running, the oil eventually reaches full working temperature and such waste products are largely evaporated.

This is why the actual time interval, rather than the mileage covered, is more important in trying to determine when the oil is due for changing; and it is the reason for recommending, in the service schedule at the end of this book, that oil should be changed on a time basis, almost regardless of mileage covered.

When badly contaminated, the old oil is still far better than none at all, but will be causing accelerated wear. The answer is the periodical draining off, preferably when the oil is still hot after a run of at least 10 miles, and replacement with fresh oil. First, identify the engine oil-drain plug, which is usually a tapered square shank nut protruding at the bottom or lower side of the engine sump, and choose a spanner which fits it correctly. Most engines hold a gallon or more of oil in the sump, and a suitable shallow container which will hold at least this quantity is needed, having a wide opening. If the drain plug is on the side of the engine, the first rush of oil will emerge in a lateral jet, not just dribble downwards.

Strongly recommended for the job of draining not only engine but gearbox oil as well, is the purchase of a 'Drainercan' made by Bell Products, of Harpenden, Herts. Shallow enough to go under the sump, it has a removable cap for emptying, and a cap and air vent in the side; it is shaped to allow the oil to run down and into the container. The main cap is

61

removed for emptying, the side is cleaned, and then the Drainercan is ready for re-use. It is obtainable from most accessory shops.

Many garages accept waste oil, or it can be disposed of by burning on the bonfire if you are not in a smokeless zone, or burying deep in the ground. Waste oil should not be poured down drains or into rivers. To emphasize the point, similar comments are made in the section dealing with draining of transmission oils.

The drain plug should be left out for about an hour, to allow the final dregs of sludge to trickle out. It is then replaced, taking care not to cross-thread the plug, and not to over-tighten. Firm tightening with a short spanner and avoiding brute force is adequate.

Naturally, fresh oil must be held in readiness before draining-off the old, and the best way is purchase in gallon tins of suitable brand and the correct grade recommended by the manufacturer. Purchase in this way from a store or accessory shop offering good value usually provides a further saving compared with the cost of having oil changed for you by a garage. Pour in the new oil through the filler orifice, checking the dipstick, until the level is up to the mark.

Filtration

Most engines have a gauze filter in the sump, which prevents any major particles of swarf or debris from being pumped through the lubrication system. This filter does not have to be serviced, but there is an additional main filter, which does most of the work of keeping the oil clean; and this *does* have to be renewed. The handbook or lubrication chart for the car should be consulted, and it will be found in most cases that replacement every 10,000 to 12,000 miles is advised. This is in line with the annual replacement recommended in the service schedule at the back of this book, but more frequent filter changing should be done if the manufacturer recommends it.

The two main types are the throwaway canister, and the renewable cartridge. The oil filter is often, but not always, low down on the left or near-side of the engine (on the right-hand as you look under the bonnet from the front); and it is usually held by a hexagonal nut. The filter change should be carried out at the same time as the engine oil is changed, because about a pint of dirty oil remains in the filter. This is normally left there, and then mixes with the new oil; but if the filter is due to be changed as well, it is logical to do so at the same time as the oil is renewed.

When the filter is removed, oil may run out, so the 'Drainer-can' should be underneath to catch it. The canister-type filter simply unscrews and is thrown away. It may take rather a struggle to free it initially, and tying a thin rope with a slip knot round it, and then binding round and round may help. You can then exert a lot of torque to start the unscrewing process by pulling sharply on the rope to turn it in an anti-clockwise direction. The cartridge variety are more easily freed, by turning the nut, first with a spanner and then quickly by hand. Its interior cartridge is thrown away and the new one put in, after first cleaning the interior of the container.

With the new filter will be found a rubber ring, which is for sealing the container to its base. First, the old ring must be removed, which may present some difficulty. A sharp pointed object such as the end of a pair of dividers will help to enable you to spear it and then persuade it out. The new ring is then pressed firmly home into the groove provided, before re-fitting the filter. If a cartridge-type filter is being changed, check the instructions with it to ensure that the new element is inserted the correct way and not upside-down.

After changing the filter and re-filling the sump with new oil, the engine should be run for a while, and the dipstick

63

re-checked. It will be found that the level has dropped
slightly as the filter compartment re-filled, and more oil may
have to be added. Also look carefully to ensure that a true seal
has been obtained between the body of the filter and the
base. Many an engine has been ruined because the oil was
pumped out of an incorrectly-fitted filter.

Driving the auxiliaries

More and more cars now have an electrically-driven fan, but
there is still a water pump and a generator – either dynamo or
alternator – to be driven by the engine. The first is to circulate
the cooling water, for without it the coolant will boil after
having driven only as little as a mile. The second is to re-
charge the battery, replacing the current taken from it by
headlamps, ignition and so on, and its output varies according
to the amount being taken out of the battery. These matters
are covered more fully in the section dealing with safety, to
which adequate electrical power is an important contributor.

Whether it also drives the fan or not, the generator and
water-pump belt is invariably called the fan belt, and its
failure is a source of many roadside breakdowns. Such trouble
can be avoided by fairly frequent inspection of the belt, since
breakage is always preceded by a stage during which the inner
part of the rubber becomes brittle and begins to crack, and the
outer side or edges to fray. These are early-warning signs, to
be taken as meaning: replace fan belt as soon as possible. Even
if they are not seen, replacement of a fan belt after three years,
as recommended in the maintenance schedule at the back, is a
sensible precaution to ward-off trouble.

A replacement belt of correct size for the make and model
of engine should be purchased, and the nuts securing the
dynamo or alternator, including the clamp nut which provides
the adjustment, should be slackened-off. The generator may

now be swung over towards the engine, making the belt very loose. Usually the belt can now be removed, feeding it over the fan blades if necessary, and perhaps levering it off the pulley gently with the aid of a wide screwdriver.

It is not unknown for there to be so little clearance between the fan blades and a surrounding cowl or the radiator that there is not room for the belt to be squeezed between the two. If this is the case, it is permissible to use a long pole such as a broomstick, gently to lever the engine rearwards on its flexible mountings to provide the extra $\frac{1}{4}$-inch clearance which may make all the difference; an assistant may be helpful here, and care should be taken to ensure that the lever bears against some strong part of the engine block, and not against one of the thin metal covers. Alternatively, it may prove necessary to slacken the nuts securing the radiator or fan cowl, re-tightening after the new belt has been squeezed through.

If there is difficulty in persuading the belt to go over the pulley, it is sometimes possible to get it as far as possible on the flange, hold it with a screwdriver or rod and keep fingers clear while an assistant gives a quick turn of the starter. Once in position, the belt must be tautened by levering the generator away from the engine and holding firmly while the adjustment clamp bolt is tightened. The remaining nuts securing the generator are then tightened as well. In correct adjustment, a fan belt should give about $\frac{3}{4}$ inch when pressed lightly but firmly at the mid-point between pulleys. After a new fan belt has been fitted, a further adjustment may be necessary following some 500 to 1000 miles' running. Subsequent adjustment may be needed three or four times in the life of the belt.

The cooling system

Although often a cause of roadside breakdown in neglected cars, the cooling system is basically trouble-free and long-suffering and requires only minimal maintenance. In a few cases, typically of course, the Volkswagen rear-engined cars, cooling is by forced air and the only attention needed is to ensure that any drive belt used is kept in good shape as already outlined. In some other cases, notably Renault, a sealed-for-life cooling system is used in which no renewal of the coolant is recommended.

For the majority of cars, with or without an overflow reservoir or catch-tank, anti-freeze is needed for two reasons; primarily, for the obvious one of preventing it from freezing in winter, and secondly for the less-appreciated need to avoid corrosion. Ordinary water in a cooling system causes rust in both engine and radiator. Anti-freeze contains inhibitors which prevent corrosion, and so it should be left in all year. The ethylene-glycol to lower the freezing point of the water remains effective almost indefinitely; it is because of deterioration of the corrosion inhibitors that annual changing of anti-freeze is recommended.

Locate the tap at the bottom of the radiator and, with the vehicle over a suitable drain, turn it anti-clockwise. If coolant does not run out, the tap may be blocked and needs to be cleared with a piece of wire. The radiator filler cap should be removed to relieve the pressure, and finally the tap (if fitted) on the side of the engine block should be opened to drain this as well. When all coolant has run out, the taps can be closed and ordinary water flushed through; fill-up, run the engine for a little while, and then drain-off again.

Finally shut the taps, and examine all the water hoses, including heater pipes. If there is sign of bulging or cracking, renewal now may save a boil-up and wasted anti-freeze later. Hose clips should be checked for tightness.

It is not often realized that for an engine to run at too low a temperature is harmful. As well as making the heater ineffective, it wastes petrol – because efficient running temperature is not reached – and it causes accelerated engine wear. If, due to poor heater performance and a very low reading on the temperature gauge, cool running is suspected, now is the time to check the thermostat.

It is usually located in a dome-shaped casting at the front of the engine, secured by two or three nuts. After undoing the nuts and perhaps giving it a light tap with a hammer to free it, the thermostat cover can be lifted off. The thermostat is of brass-coloured metal comprising a disc and a surround. It is operated by a small container of wax or a bellows containing liquid beneath the disc. All that needs to be confirmed is that the disc is firmly seated against the surround. If it is standing clear by any substantial amount, the thermostat is faulty since it should open only when surrounded by water at the opening temperature – usually 85 deg. C, only 15 deg. C off boiling. There is no repair possible: the faulty thermostat must be replaced, and it may be advisable to fit a new housing gasket at the same time.

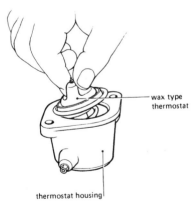

wax type thermostat

thermostat housing

A wax-type thermostat.

When re-fitting, ensure that the thermostat has properly seated itself with its rim in the surround, before re-fitting the housing and tightening, otherwise the housing itself can easily be cracked. Thermostats usually fail in the open position as described. For them to fail closed is less likely, and results in seemingly inexplicable overheating.

When finally satisfied with the condition of the cooling system, consult the handbook to see what the coolant capacity is, and hence how much anti-freeze will be needed. Purchase in 1- or 2-pint tins is advisable and any surplus can be retained for subsequent topping-up if needed. Pour in about a pint of water, followed by the appropriate quantity of anti-freeze as directed by the manufacturer and according to the degree of frost protection required; and finally top-up with water. After a run of moderate length the anti-freeze will be thoroughly mixed through the system. Keep an eye on the coolant level for the next few days after changing the anti-freeze; further topping-up may be necessary as airlocks are dispersed.

The perfect spark

For an engine to run, it requires a petrol-air mixture of correct proportions, sparked off by the ignition system at the correct time. The nature of the petrol engine is such that it is forgiving, and will still run after a fashion if these requirements are imperfectly met; i.e. an over-rich mixture given an inadequate spark either too late or too early, may still allow the engine to run, albeit inefficiently. As well as doing itself harm because of the incomplete combustion, an engine in this state gives less power than it should and returns poor mpg figures. Of the two faults, the ignition is far more likely to give trouble than the carburation, and so it is with the ignition that we begin, in search of the perfect spark.

When sparking plugs are tested, they are screwed into a

machine which simulates the pressure developed in the engine, and the pressure is increased until the point is reached when the current finds it easier to jump across a test gap incorporated in the machine instead of across the gap in the sparking plug. This is called the pressure at which the plug begins to break down, and is a measure of its efficiency. A garage mechanic with such a machine at his disposal can tell by measurement whether a given sparking plug is still within limits or not. The home mechanic has no such equipment at his disposal, and must hope for the best.

However, if it is appreciated that sparking plugs do *not* last for ever, that they do become inefficient in extended service even though they may still look all right, we are half-way to successful engine maintenance. Annual replacement on a presumed mileage of 10 to 12,000 is recommended, and at approximately midway, i.e., every 6 months, it is advisable to clean and adjust the sparking plugs. To do this properly it is necessary to use a sand-blasting plug cleaner, which may mean taking the plugs round to your local friendly garage and seeking the loan of their machine for a few minutes.

After blasting, the plugs are re-set to the correct gap as recommended by the car manufacturer. A feeler gauge (purchased from any accessory store) is used, and if the recommended gap is 0·025-inch (a typical figure, called by a mechanic 25 thou), the '15-' and '10-thou' feeler gauges are put together and used to assess the gap between the two electrodes. If it is too loose, the lower electrode is tapped lightly to bend it inwards slightly; if too tight it is gently levered the other way with a screwdriver, bearing against the surround and not the centre electrode. Adjustment to give a firm sliding fit as the feeler gauge is pressed between the electrodes soon shows that the correct gap has been obtained. Plugs are tested on the machine *after* gapping, and when finally checked they should be blown out with the air line to ensure that no sand from the cleaning process remains inside, which could be dislodged and damage the engine. Bear in mind that even new plugs need to have the gap adjusted before fitting; they will not necessarily arrive with the gap set as recommended for your car. Only the make and grade of plug recommended by the car manufacturer may be fitted to an engine. When screwing a plug in, do not over-tighten; a good guide is to use only one hand, and to have the tommy bar in the central or 'T'-position, to reduce the available leverage. Check that the sealing washer is not omitted, or the former washer inadvertently left in the orifice in addition to the new one.

The distributor

Given the right electricity at the right time, the sparking plug will give the perfect spark we are seeking for efficient engine performance, if it is within its approximate 12,000-mile life and has been correctly cleaned and gapped. The timing of the spark will be dealt with later; here we are concerned with the surprisingly simple mechanism by means of which the high-voltage current is provided and fed to the plug. Although the design of the distributor – the key component in the ignition system – varies from one make to another, the principle is the same. Only in the Jaguar twelve-cylinder engine, at the time of writing, is a completely different system of transistorized, sealed-for-life electronic ignition used.

The basic principle of ignition makes use of the fact that electricity takes the easiest route to complete its circuit. Imagine the centre of London as the battery, and London Airport as the destination; let one wire run there direct, and the other wire run there via Birmingham and the Severn Bridge. Obviously, the current will take the shorter route; but if, while it is flowing the easy way, this short wire is

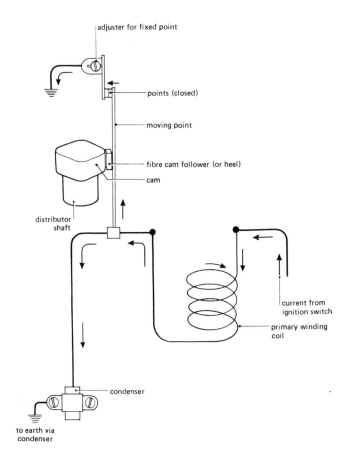

adjuster for fixed point

points (closed)

moving point

fibre cam follower (or heel)

cam

distributor shaft

current from ignition switch

primary winding coil

condenser

to earth via condenser

Schematic diagram of ignition circuit:
a. points closed.

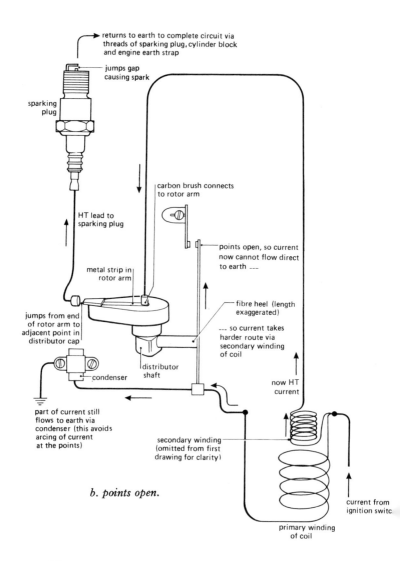

returns to earth to complete circuit via
threads of sparking plug, cylinder block
and engine earth strap

jumps gap
causing spark

sparking
plug

carbon brush connects
to rotor arm

HT lead to
sparking plug

points open, so current
now cannot flow direct
to earth ---

metal strip in
rotor arm

fibre heel (length
exaggerated)

jumps from end
of rotor arm to
adjacent point in
distributor cap

--- so current takes
harder route via
secondary winding
of coil

condenser

distributor
shaft

now HT
current

part of current still
flows to earth via
condenser (this avoids
arcing of current
at the points)

secondary winding
(omitted from first
drawing for clarity)

b. points open.

current from
ignition switc

primary winding
of coil

suddenly cut or interrupted, it must now take the long route round. Now imagine the wires not spread out across the country, but wound in two insulated coils round a bar of soft iron. As the current flows through the short coil, which is of thick, low-resistance wire, it induces a magnetic field around the iron centre of the coil. When this is interrupted, ·the sudden collapse of the magnetic field induces a very high voltage flash of current in the second coil, which is of much longer, thinner wire. This is the high-voltage coil which is taken along thick leads to the sparking plugs.

Fed into the centre electrode of the plug, it jumps the gap to the outer electrode, making the vital spark which fires the mixture in the process, and then returns to earth through the cylinder block of the engine and the steel body of the car.

The distributor's chief function is the repeated making and breaking of the short (or primary) circuit, which it does by opening and closing little switches called the distributor points. The switch is worked by a square shaft geared to the engine

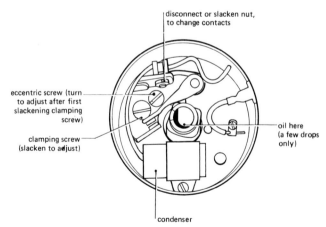

Screw adjustment for contact points gap.

73

and so arranged that as the piston comes to the top of its stroke compressing the mixture, one of the squared corners of the shaft pushes the points apart. The current now has to go the long way round and jump the gap in the sparking plug, igniting the mixture. In a six-cylinder engine a hexagonal shaft must be used to provide the additional number of sparks.

A further function of the distributor is to send the current from the coil to the correct sparking plug in the right sequence,

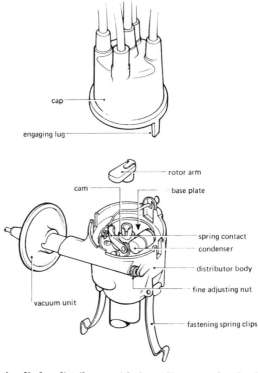

Lucas 4-cylinder distributor with fine adjustment thumb wheel – often omitted.

74

called the firing order. This is simply achieved by feeding the sparking-plug leads to equidistant points inside a circular cap, and contriving that a metal bridge (called the rotor arm) is in position to carry the current from the centre to the appropriate terminal at the critical fraction of a second when the current flows from the coil. For convenience, the same shaft which works the switch or points also carries the rotating bridge or rotor arm.

There are further refinements in the distributor. A condenser is wired-in as a sort of by-pass of the points, which has the effect of lessening arcing across the points, extending their life. Because the spark needs to occur in each cylinder progressively earlier, within limits, as the speed of the engine increases, a system of centrifugal weights is built into the base of the distributor (or, in some A.C.-Delco models, it is at the top). It is not easy to gear the shaft, which opens and closes the points and turns the rotor arm, to rotate at any but a speed in fixed proportion with engine speed. So an extension shaft is fitted at the top, which is turned through several degrees in the direction of rotation as the engine speed increases. This is called the centrifugal advance mechanism,

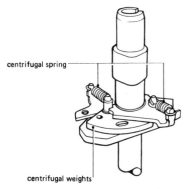

centrifugal spring

centrifugal weights

Centrifugal advance mechanism in a distributor.

and it is this lightly spring-loaded movement which makes the rotor arm and shaft seem loose if you try to turn them with the engine at rest.

Further adjustment of the time (relative to piston position) at which each spark occurs is provided by the vacuum advance mechanism which is connected by a small-bore pipe to the engine inlet manifold. This makes the necessary compensation for the fact that the spark needs to occur earlier at tickover or when the engine is running on small throttle openings than when the driver has his foot hard down on the accelerator. This is the function of the bell-shaped pressing at the side of the distributor.

Distributor maintenance

In service, the contact points of the distributor which switch the sparks on at the rate of several thousand a minute at the

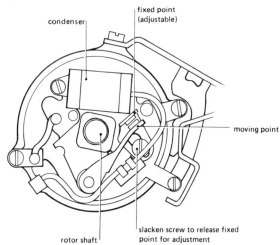

The more usual type of distributor has no fine adjustment provision for setting the gap.

76

upper engine speeds, gradually erode. They finally reach a stage where a substantial pimple will have formed on one and a corresponding crater on the other. Replacement points are available for every engine, and in comparison with the costly waste which can result from using worn-out points, they are not expensive. Without proper apparatus it is impractical to renovate distributor points; and again, replacement is easier.

In the schedule at the back, renewal every year is advised

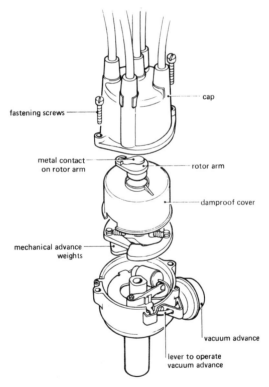

fastening screws

cap

metal contact on rotor arm

rotor arm

dampproof cover

mechanical advance weights

vacuum advance

lever to operate vacuum advance

In the AC Delco distributor the centrifugal advance mechanism is above the contacts and there may be a damp proof cover.

77

but it is emphasized that this is nearing the limit for normal point life if some 12,000 miles will have been covered. So inspection should be carried out at 6,000-mile intervals, and if the points show signs of pitting, replacement then is advisable.

The job is easily carried out. First, remove the cap of the distributor. This is usually done by snapping-off a flat metal spring at either side, but on some Delco units (Opel and Vauxhall), the cap may be secured by a cross-head screw on each side. A new set of points of the correct type for the engine will have been purchased, and a glance at the contents of the pack will show even a complete novice what he is looking for: the points take the form of a fixed component held by one or two vertical screws, which also provide the adjustment, and a long spring section with a fibre heel bearing against the squared shank in the middle of the distributor. This component is located usually by a horizontal screw, and should be removed first. In some layouts it is held at the end by a nut and washer, which should be undone, taking care that no screws or nuts are allowed to fall down into the base of the distributor. Note the arrangement of wires.

Now lift off the fixed component after removing the locating and adjusting screw – again taking care not to drop screws. New points should be cleaned with a little petrol to remove the protective coating. Fit the new components in the reverse order – short fixed component first, with the screw just lightly tightened; then, the spring section, taking care that the wires – one connected to the condenser and the other to bring the low tension current to the contact point – are re-positioned correctly.

The engine must now be turned by hand. If a starting handle is provided, there is no problem. If not, and the car has manual transmission, the brake should be released (car on level surface) and top gear selected; the engine can now be turned by gently pushing the car. If it has automatic transmission, one solution is repeated momentary use of the starter; the other, and preferable one, is removal of the sparking plugs, after which it should be possible to turn the engine by pulling on the fan belt. Whichever method is used, the engine is now turned slowly to a point where the fibre heel of the moving contact point is on the very top of one of the four (in a four-cylinder engine) angles or cams of the distributor shaft.

Now check the handbook to see what distributor gap is recommended for the engine – a gap of 0·015-inch (or 15 thou) is typical. Select feeler-gauge blade or blades to this value and wipe clean before offering them between the points. The fixed component is now moved towards the spring-loaded cam-operated one until the gap just allows light sliding contact with the feeler gauge. Tighten the adjusting screw and check again,

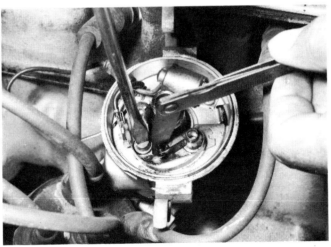

A screwdriver is used to loosen the adjustable point, and the gap is checked with the feeler gauge. The gap is correct when the feeler gauge moves between the points with slight 'drag'. The gap should be re-checked after tightening the locking screw.

re-adjusting if necessary. The moving point should not be seen to spring inwards as the gauge is removed; if it does, it shows that the feeler blade is holding the points apart and that the gap is too small. Now gently turn the engine in either direction; if the gap increases, the heel of the point was not on the peak of the cam, and adjustment must be done again.

Following the fitting of new contact points, the static setting of the ignition timing must be adjusted (see Ignition Timing). It is also advisable to re-check the gap after a few hundred miles of running. Due to initial wear and bedding down of the fibre heel, the gap may close slightly and have to be re-set, followed again by re-adjustment of the ignition timing.

In addition to the renewal of points, two or three drops of thin oil should be dribbled down the centre of the distributor drive shaft after removal of the rotor arm (this just pulls off). A light smear of grease should be applied to the cams or squared shank of the distributor shaft. Lubrication should not be overdone, as if oil finds its way on to the contact points, the engine will not run. These attentions are done more often than point renewal, and at the same time a visual check should be carried out: observe that the carbon brush which carries the high tension spark from the centre terminal to the rotor arm is in place, examine the points for pitting, look out for cracks inside the distributor cap moulding, and wipe clean with a dry cloth. If a crack is present, tracking of the high tension current and misfiring will result; the cap will have to be replaced.

Firing order

When removing the distributor cap there is a possibility that one or other of the high-tension leads to the sparking plugs will pull out, especially if suppressors to reduce radio and television interference are incorporated in the lead. Correct re-positioning is important. Similarly, if the sparking-plug leads are pulled off the plugs for a plug change or cleaning and gapping, re-positioning in the correct order is vital. This is such an important, basic statistic of the engine that it is usually given in the handbook, and it is often stamped on the rocker box or manifold. On German cars it is preceded by the word *Zundfolge*.

Most four-cylinder cars fire in the order 1 – 3 – 4 – 2; the numbering is always from the front of the car rearwards, or in the case of a transverse engine it is from the pulley end towards the transmission or clutch end. Some four-cylinder engines are designed to fire in the order 1 – 2 – 4 – 3. What you *never* find is the order 1 – 2 – 3 – 4, which would give vibration and terribly lumpy running. A six-cylinder engine usually fires in the order 1 – 5 – 3 – 6 – 2 – 4; a less-often used alternative is 1 – 4 – 2 – 6 – 3 – 5.

If in doubt about the order it is a good idea to label the leads with numbers under a band of Sellotape before removing them. Some will be of such length that they could not be confused, but there will often be some leads which could equally well be re-fitted to the wrong plug.

Ignition timing

As explained in the section on the distributor, provision is made for automatically adjusting the timing of the spark in each cylinder, making it progressively earlier when the engine speed is high, or when throttle openings are small. Earlier than what? The answer is the basic setting of engine ignition timing, governed by the positioning of the distributor. The main drive shaft of the distributor is positively geared to the engine, and turns usually – but certainly *not* always – anti-clockwise. If the distributor is turned a few degrees the same way as the direction of rotation of the rotor arm this makes the timing later; it is called 'retarding the spark.' Turning the other way, backwards against the direction of rotation of the shaft, it makes everything happen earlier and is called 'advancing the spark'.

We are now dealing with a critical aspect of engine performance, and careful work is called for. Quite a small movement of the distributor away from the correct position can be sufficient to make the engine scarcely run at all, so it can be visualized that a minor error will still leave it badly down on performance and economy. Because the standard timing is affected by the exact positioning of the contact-breaker points, re-adjustment is necessary, as mentioned earlier, after new points have been fitted or the gap altered. There are two ways of setting the timing, and we deal first with what is called the static method.

Before anything else, we must identify the timing mark, and know whether it is a timing mark or a TDC (top dead centre) mark. A timing mark is one to which the distributor is to be set; on some cars it may just be a small notch in the flywheel.

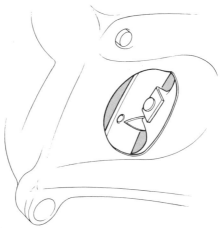

Typical timing mark on a flywheel, seen through a 'window' in the housing.

A TDC mark identifies the position (in terms of rotation) of the crankshaft when the piston in number-one cylinder is at the highest point of its travel. This is *not* the position to which the distributor is set, as it usually requires the spark to be a fixed amount (perhaps 10 deg.) before the TDC position.

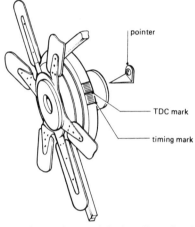

Ignition marks may be on the crankshaft pulley; the timing mark must be identified for ignition setting – not the TDC mark.

What does all this mean in practice? First, one must consult the handbook to see what is said about engine timing and the position of the mark. In the absence of information, one must simply look for it. A notch in the lower pulley of the engine, or a series of lines with a zero mark on the flange of the crankshaft damper (like a large metal disc behind the crankshaft pulley) are typical positions for the timing mark. Look also for a pointer nearby (usually above). On some engines a little inspection hole, perhaps with a swivel-back cover, is provided in the flywheel housing at the back of the engine, and the timing mark is on the flywheel itself. This is less convenient, as the mark can be seen only when it is behind the window. To help find it, remove the distributor cap, noting first the approximate position of the terminal connected to number-one sparking plug. Now turn the engine until the rotor arm points to the position where this terminal was located.

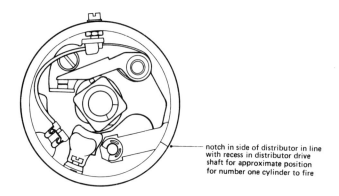

notch in side of distributor in line with recess in distributor drive shaft for approximate position for number one cylinder to fire

Once the mark has been located and its pointer identified, the engine must be rotated (see Distributor Maintenance) until the mark and the pointer are exactly in line. If the mark is a TDC mark rather than a timing mark, a position providing the correct number of degrees of advance before TDC must be obtained. There is usually easy provision for this by such means as two pointers, or two marks – one a TDC mark and one a timing mark. Again, the handbook or information from a garage or the manufacturer is necessary to locate the correct position for timing.

Now identify the thin wire – not the thick HT lead – which runs from the coil to the distributor. Take a 12-volt inspection lamp or test probe, and connect one end to a suitable earth such as the battery earth lead if within convenient reach. The other wire from the inspection lamp is connected to the terminal at either end of the coil-distributor wire. Now remove the distributor cap and slacken the clamp bolt at the base; and switch on the ignition. Leave the car in gear as it will dis-

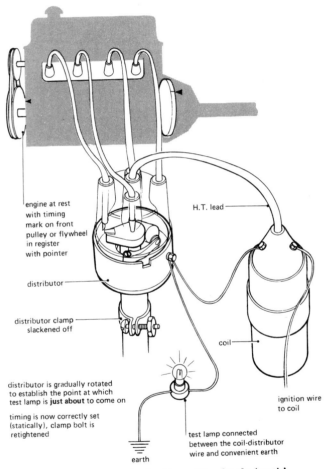

engine at rest
with timing
mark on front
pulley or flywheel
in register
with pointer

H.T. lead

distributor

distributor clamp
slackened off

coil

distributor is gradually rotated
to establish the point at which
test lamp is **just about** to come on

timing is now correctly set
(statically), clamp bolt is
retightened

ignition wire
to coil

test lamp connected
between the coil-distributor
wire and convenient earth

earth

*Using a circuit test lamp for ignition setting by the 'static'
method.*

courage the engine from turning accidentally, and it cannot start with the distributor cap removed.

The body of the distributor can now be gently revolved; only a fraction of movement is necessary – aim for a ¼-inch in either direction at the most. It will be found that as the distributor is rotated against the direction of rotation when running, advancing the ignition, the lamp comes on, showing that the points have opened; turned the other way, the points close and the lamp goes out. By experiment a position should be found where the lamp is out, but just on the point of coming on. Check by turning gently that the rotor arm is in the fully-back position, i.e. that no movement of the centrifugal advance mechanism is giving a false reading. The slightest turn of the rotor arm in an anti-clockwise direction (or the same direction as that when turned by the engine) should now light the lamp, confirming that the points are just on the verge of opening.

The distributor is now correctly set. Check that the timing mark and pointer are still in position, and tighten the clamp bolt; re-check that the distributor is still set correctly (sometimes it can move when re-tightening the clamp bolt). Replace the distributor cap, remove the test lamp and switch-off the ignition.

Stroboscopic Timing

It is generally considered that timing with a stroboscopic lamp is preferable to the system previously outlined. A special neon strobe must be acquired; good ones are now obtainable from accessory shops at moderate prices – a long-lasting acquisition to which every home mechanic sooner or later must be treated. Again the timing mark – not the TDC mark – must be identified, and it helps to put a spot of white paint or a chalk line on it. With the engine running at a slow tick-over, the light is aimed at the mark, and gives an instan-

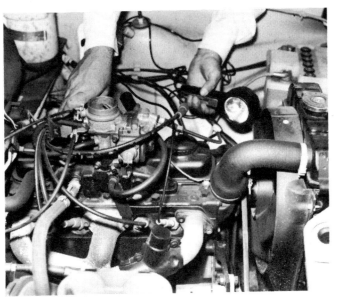

An ignition timing light, such as this KLG one, enables accurate adjustment of the distributor to be made against the timing mark, with the engine ticking over. It is important to know what mark is to be used at tickover, and on what sparking plug lead the engine timing should be set. Some badly located timing marks are very difficult to reach with a timing light, and one must then resort to the 'static timing' method.

taneous flash at the precise fraction of a second when the spark occurs in number one cylinder, to whose sparking plug it is connected. Again the distributor clamp bolt is slackened, the distributor gently rotated bodily either way as required to put the timing mark, lit up by the strobe light, precisely on the pointer.

Two points to note are that the vacuum pipe from the inlet manifold to the distributor should be temporarily disconnected when timing this way, to prevent vacuum advance from giving a false reading; and that the manufacturer's information should be checked carefully to see what stroboscopic timing is recommended. It may well differ from that advised for the static method.

On some engines such as that of the Opel Commodore, the recommendation is to use a stroboscopic lamp, but it is done with vacuum pipe disconnected, all sparking plugs removed, and the engine turned over on the starter.

Timing lights often give a rather weak light, and it may help to run the front of the car into the garage; but ensure good ventilation, and exhaust feeding out of the garage, when adjusting timing with the engine running.

Carburation

While ignition will almost certainly become inefficient if left unattended for an extended mileage, carburation is equally likely to go on giving adequate service and with minimal need for maintenance. Some attention is called for, however, particularly in the needle-in-jet carburettors ('S.U.' and 'Stromberg') dealt with later. On what are called fixed-jet carburettors, as used on most Fords, for example, there is little call for routine attention, and in fact no provision for altering the mixture for ordinary running at all. Only the slow-running mixture and tick-over speed can be altered.

It used to be simple to identify two screws, one affecting the closed position of the butterfly, and the other – the mixture adjustment screw – having a half-slot in its head to serve as a pointer and discourage use of a screwdriver; and on some cars these adjusters can still be found. Unfortunately, with greater importance being attached to the reduction of exhaust pollution, more and more cars are being turned out with sealed or tamper-proof emission-control carburettors. The intention is

88

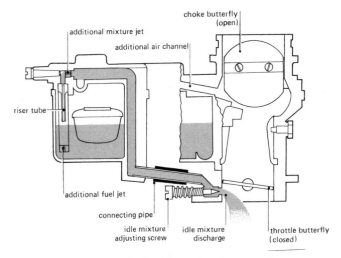

Sectioned view of a simple fixed jet carburettor.

that these should be re-set only with the car connected to an exhaust gas analyser, and is so out of scope of the average owner-mechanic.

As explained in the service schedule at the back of the book, carburettor adjustment should not be attempted until it has been established that the ignition is in good order; otherwise time will be wasted in wrongly adjusting carburation to compensate for ignition faults. Further sound advice, often overlooked by service-station mechanics, is that the engine must be thoroughly warmed through by a run of at least 5 miles before adjusting the carburettor. If you are in doubt as to the make of carburettor fitted to your car, the handbook or a road test of the model should elucidate; or visual inspection may reveal a maker's name moulded into the casting of its body.

'Zenith', 'Solex' and 'Weber' Carburettors

Being basically similar, these carburettors can be dealt with together. 'Solex' and 'Zenith' carburettors have an economy device to weaken the mixture of part throttle operation by admitting extra air ('Zenith') or by enriching a weak mixture on wide throttle openings ('Solex'). Adjustment is provided only for the tick-over, and it is important to appreciate that setting of the slow-running adjustment (both mixture and volume) does not influence the normal behaviour of the carburettor under load. In each case there should be a throttle-stop-screw for adjustment of volume, and an additional screw (usually identifiable by the presence of a spring behind it to prevent accidental rotation) which adjusts mixture. Screwing-in usually – but not always – weakens the mixture.

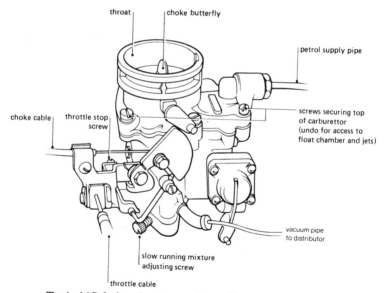

Typical 'Solex' carburettor to show adjustment screws.

90

With the engine ticking over and, if possible, with the air filter fitted, turn the throttle stop-screw (usually anti-clockwise) to reduce the tick-over speed to the minimum without reaching the stage of stalling. Now turn the mixture screw first in then out, noting that at each extreme the engine begins to stagger and will stall if allowed to. Between these extremes will be an ideal giving the fastest tick-over speed without re-setting the throttle stop-screw. Once this position has been established, leave the mixture screw alone, and adjust the throttle stop-screw to give a steady, vigorous but not too fast tick-over. Quickly flick the throttle open and shut; the

*Setting the slow running adjustment on a 'Solex' carburettor.
Linkages and fastening screws often obscure the important
mixture and slow running adjustment screws, and it may
help to identify them by having an assistant work the
accelerator while you watch to identify the movement of the
throttle connection. The mixture adjustment is often just a
random-location screw, usually with milled surround.*

91

engine should zoom and settle back to tick-over again without any stalling.

Extreme difficulty in getting any sort of reasonable tick-over suggests a blocked jet, and being the smallest, the idling jet is the most likely to be at fault. On the 'Zenith' carburettor the float chamber is detachable. Look for a series of screws flush with the upper casting, and carefully undo them. This frees the float chamber, which comes away full of petrol; take care not to tear the thin gasket which makes the seal between the two castings. Pour away the petrol and gently tip the brass-coloured float out of the chamber. Several brass 'screws' will now be seen in the float chamber casting, and they should be unscrewed one at a time and inspected. Holding them up to the light, look for a pinhole of light coming through, and if there is an apparent blockage, clear it by blowing through. Do not be tempted to poke anything through the tiny orifice, except – as a last resort – a clean bristle of a toothbrush. When re-assembling, check that the float is inserted the correct way up, and that the float chamber seats firmly against the gasket before any attempt is made to tighten the screws. Tighten-up progressively, and without violence. On the 'Solex' and 'Weber', the float chamber remains *in situ*, and the top cover is removed by unscrewing.

If a manual choke is fitted, check that when an assistant operates the choke control the choke butterfly goes fully to the off position when the knob is pushed in. Use the oil-can to lubricate accelerator linkage joints or the end of the throttle cable, but do not put oil on the actual spindle of the throttle butterfly in the carburettor; it is intended to be lubricated by the petrol vapour passing the spindle.

Twin 'Zenith' carburettors have been fitted on some engines, notably early examples of the Vauxhall VX4/90. Adjustment is rather more complicated, as the linkage between the carburettors must be slackened-off, and adjustments

92

Adjustment of this modern twin-choke 'Weber' carburettor for the V6 3-litre Ford engine is straightforward: the screwdriver is being used to adjust the throttle stop screw, setting the idling speed (or intake volume). The mixture control screw is in view at the base of the carburettor on this side, to the right of the sparking plug leads.

carried out individually; but with a little practice one can soon identify the point where one carburettor begins to open more than the other, and thus to balance them. In wider use is the twin-choke unit, such as the twin 'Solex' and 'Weber'. These are essentially two carburettors in one, and one barrel only is used in light throttle work. Adjustment is as for the single unit, and is easier than with separate carburettors.

93

'S.U.' and 'Zenith-Stromberg' carburettors

These are broadly-termed the needle-in-jet carburettors, and differ from those described earlier in that the proportioning of fuel to air is metered by the progressive withdrawal of a tapered needle in a single jet. The higher the needle is lifted, in relation to the position of the jet, the more petrol can pass through the orifice, and the carburettor relates this positioning of the needle to the amount of air being delivered, as determined by the driver's foot on the accelerator. For cold starting, instead of closing an additional butterfly to reduce the air flow, the jet is temporarily lowered, thus enriching the mixture. This is why the cold start device for an 'S.U.' or 'Zenith-Stromberg' carburettor is called the mixture control, rather than the choke.

The important difference of these from the multi-jet carburettors is that the mixture adjustment for idling speed also affects the mixture delivered to the engine throughout the speed range. Correct adjustment is therefore particularly important. As a routine check, an easy test can sometimes be made by locating a small pin or toggle which protrudes down at the side of the carburettor, just beneath the bell-shaped top cover. If this is pushed upwards while the engine is ticking over, it has the effect of temporarily weakening the mixture because it raises the air valve slightly. If the engine speed increases during this test, the indication is that the mixture is too rich; if it increases momentarily and then settles down to run at roughly the same speed as before, the suggestion is that the mixture strength is about right. Too lean a mixture is indicated by stalling when the lifting pin is raised. Some carburettors, such as the 'S.U.' on the Allegro 1500 engine, no longer have this helpful means for checking mixture.

Both the 'S.U.' and the 'Stromberg' have an internal damper, which prevents too sudden movement of the air

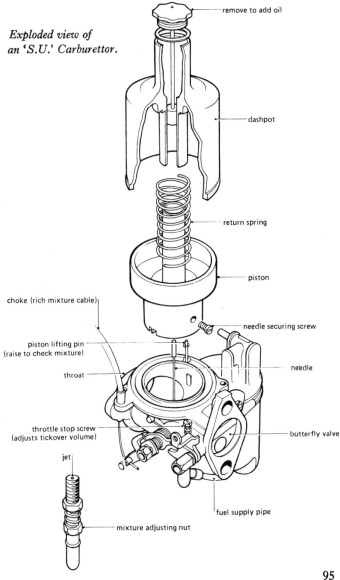

*Exploded view of
an 'S.U.' Carburettor.*

remove to add oil

dashpot

return spring

piston

choke (rich mixture cable)

needle securing screw

piston lifting pin
(raise to check mixture)

throat

needle

throttle stop screw
(adjusts tickover volume)

butterfly valve

jet

fuel supply pipe

mixture adjusting nut

95

valve, providing the extra enrichment for acceleration, and a routine service requirement is to unscrew the nut at the very top of the carburettor and withdraw it complete with the short rod attached; then pour in engine oil until the centre part of the carburettor is practically full.

If the mixture check outlined suggests that the carburettor is set too rich or too lean, the jet must be raised or lowered accordingly. Some 'S.U.s' have a remote screw for mixture adjustment; the Rover 3-litre was an example. A more usual arrangement is for the jet to be raised or lowered by turning a brass-coloured nut or screw at the base of the carburettor. Err on the side of leanness when adjusting, but do not overdo it. After adjusting the mixture, the throttle stop-screw must be re-set as for the multi-jet carburettors.

Many 'Stromberg' carburettors – as on Hillmans post-1973, for example – are what is called "tamper-proof". It will be found that there is no mixture adjusting nut or screw beneath the carburettor. The jet is fixed, and mixture adjustment is made by raising or lowering the needle from above using a special tool inserted into the dashpot after removal of the cap. This intentionally leaves no scope for the owner to adjust the mixture setting, unless he is able to obtain the special tool needed for the job.

With the engine switched off, ask an assistant to pull the cold starting mixture control, and it will be noticed that as well as lowering the jet, this moves a lever bearing against the throttle linkage with screw adjustment intended to open the throttle slightly when the control is pulled. This screw should be set to give about 1 mm clearance when the mixture control

The basically similar 'Zenith-Stromberg' carburettor. More and more cars now omit the easy mixture adjustment shown here.

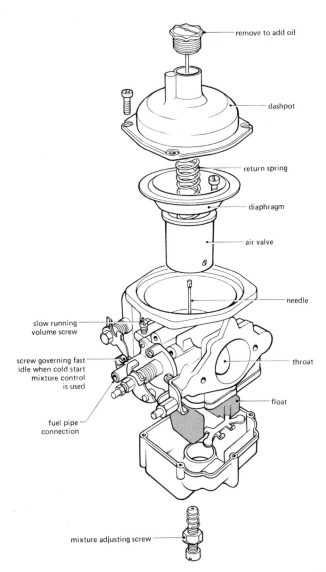

remove to add oil

dashpot

return spring

diaphragm

air valve

needle

slow running
volume screw

throat

screw governing fast
idle when cold start
mixture control
is used

float

fuel pipe
connection

mixture adjusting screw

97

is pushed fully home, and care should be taken not to mistake it for a slow-running control screw.

Centring the jet

With the air filter removed, the vertical movement of the piston to which the needle is attached can be observed, and a periodic check should be made to ensure that the jet is correctly positioned in relation to the needle. This is done by unscrewing and removing the nut and rod at the top of the carburettor dashpot, and gently lifting the piston; the piston-lifting pin mentioned earlier can be used. On releasing, the piston should fall back smartly with an audible clonk as it comes to rest against the jet bridge.

If its return is slow or there is a tendency to stick, we must do what is called centring the jet. The principle is the same for 'S.U.' and 'Stromberg' – simply that the jet is released and screwed-up to its limit, and the needle then lowered into it to set the position of the jet, which is then re-tightened. On the 'S.U.' series 'H' carburettors, the spring between the jet locking nut and the jet adjusting nut is removed, and the jet is raised as far as possible by turning the jet adjusting nut while holding the piston up to keep the needle clear. On the 'HS' series of 'S.U.' carburettors, the spring and the adjusting nut are removed for jet centring. With the 'Stromberg', the jet is held between a housing and the vertical jet-adjusting screw; a spanner is used to slacken-off the housing nut, and the adjusting screw is turned to raise the jet to the level of the bridge – it will be seen through the mouth of the carburettor. Again, the needle attached to the piston is lowered into it to locate the jet, and the housing is then re-tightened. In each case, the central nut and rod at the top of the domed dashpot should be removed for the procedure, and a check should be made for free movement of the piston after the jet has been re-tightened.

Renewing the 'Stromberg' diaphragm

On the 'Stromberg' carburettor, a neoprene diaphragm is used between the piston and the dome-shaped top cover. Under continual attack by petrol and oil, this eventually becomes distorted and may swell and even tear; and replacement is recommended as part of the carburettor service every second year. The four locating screws securing the cover should be removed, and the top lifted clear. The piston and its needle may now be carefully lifted out complete with the diaphragm. Taking great care not to bend the needle, undo the four screws which clamp the top plate to the piston, and fit a replacement diaphragm. Note that the diaphragm has a small tongue at one point; this should face downwards, and engage with a slot in the body of the carburettor. Diaphragm exchange can be done without lifting the piston out, but removal to the bench lessens the risk of straining and bending the needle. Re-centring the jet afterwards is advised.

Aids to carburettor tuning

Mixture adjustment by means of the needle-lifting pin is a rather hit-and-miss method, and as explained, correct setting in needle-in-jet carburettors is important, since this affects operation throughout the speed range. Much more accurate setting can be achieved using the Gunson Colortune 500 kit, an inexpensive and very helpful aid for the amateur mechanic. Produced by Gunsons Colorplugs Ltd, of 66 Royal Mint Street, London E1 8BR, it is available from all good accessory shops.

The special plug supplied has a transparent top, through which combustion can be observed. The 'Colortune' plug is screwed-in to take the place of one of the sparking plugs, and connected to the ignition. A plastic tube with mirror is supplied, called the 'Viewerscope'. As the mixture adjusting

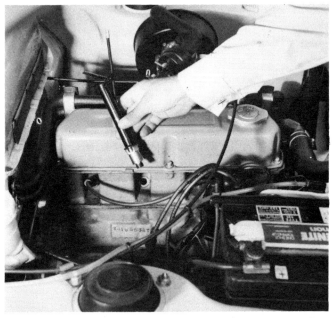

Preparing to use a Gunsons Colorplug on a Hillman Avenger. The device takes the place of the conventional plug, and its h.t. lead protrudes through a slot in the side of the "mirror tube". This lead is connected to the h.t. lead, and while the engine is ticking over the flame in the engine can be observed via the "mirror" in the top of the tube. Mixture adjustment is made to give the correct shade of bunsen blue flame in the engine.

screw or nut on the carburettor is turned, with the engine ticking over, a pronounced change in colour of the flame will be seen. When the carburettor is delivering too rich a mixture, the flame will be decidedly orange, and will change to a pale whitish-blue when the mixture is too weak. The correct setting is the strong bunsen-blue colour obtained on weakening from the over-rich state indicated by the orange flame. The transition to bunsen-blue is quite abrupt and thus enables the correct fuel-air mixture to be obtained very precisely.

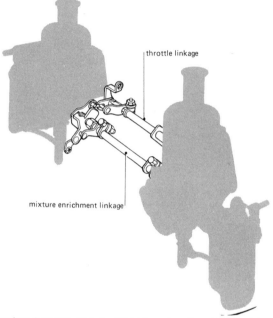

throttle linkage

mixture enrichment linkage

Connecting linkages on twin 'S.U.' carburettors for throttle and cold starting mixture enrichment.

Multiple carburettors

A few high-performance cars such as the Jaguar 'E'-Type in its earlier six-cylinder form, had three 'S.U.' carburettors; but this is rare. Many engines, such as the Triumph 2500TC, have twin 'S.U.' carburettors, or there may be twin 'Strombergs', as on the Humber Sceptre. With a single-carburettor engine, tuning is easy – first set the mixture, then adjust the

101

throttle opening to give a suitable even tick-over. With twin carburettors, the mixture delivered by each one must first be adjusted separately; then the clamps on the rod connecting the throttles of the two carburettors must be slackened off, and the butterfly screws (slow running screws) must be adjusted individually to give the correct *volume* of fuel-air mixture.

This can be done by using a piece of pipe as a form of stethoscope, to listen to the hiss or rush of air going into the carburettor mouth. Again this is not a very reliable method and a carburettor balancing instrument is invaluable for

Using a Gunsons Carbalancer to set the idling screw adjustment on 'S.U.' carburettors of a Triumph Dolomite Sprint. Suction from the carburettor throat is indicated by position of a pointer on the Carbalancer scale. Idling (volume) screws are adjusted so that the same reading is obtained with each carburettor. The actual reading is unimportant provided the tickover speed is about right, and that the reading is the same *for each carburettor. After adjustment, the clamping nuts on the throttle connections should be retightened, and adjustment checked.*

accurate tuning of a twin-carburettor engine. The manufacturers of 'Colorplugs' also produce the 'Carbalancer' for
this purpose: a rubber cone is held against the mouth of the
carburettor, with the air filter removed and engine ticking over.
The vertical position of a pointer, rising and falling in a glass
tube against a scale, shows the amount of suction of that
carburettor. The two carburettor throttles can now easily be
adjusted to deliver equal volume, and then the connecting
rod is re-clamped.

When adjusting twin carburettors, take care that movement
of one carburettor's throttle does not slightly open the other,
even with the connecting rod slackened off. Aim for a reasonably lively but not fast-running tickover – about 600 to 800
rpm if a rev. counter is fitted.

Air filters

Filtration of the air for combustion in an engine used to be by
means of an oily wire gauze, but much more refined paper

*An air filter change is easily done on most cars, and removal of a
blocked paper element filter and replacement with a new cartridge
can often make a big improvement in efficiency and economy.*

103

elements are now used. A vast variety of boxes and containers is used for the filter, differing on almost every car, some fastened by screws and some by clips. If screws are used, take care that a screw is not allowed to fall into the carburettor mouth; fortunately, captive screws are normally used. The filter box should be opened annually, and the paper element taken out and discarded; a new one is inserted and the container re-fastened. It will be found that in many cases there is no need to remove the filter box from the engine.

Winter and summer provisions

Some multi-jet carburettors have an accelerator pump, designed to squirt extra fuel into the throat of the carburettor for acceleration. The linkage is often provided with two or even three connecting holes, and for winter operation the

Most cars nowadays have provision for admitting ambient air in summer, and air heated from the exhaust hot spot in winter. The seasonal change is usually easy to do, as on this Chrysler 2-litre.

104

connecting link should be taken out of the upper hole and re-positioned in the next one down. If three holes are provided, it is not normally necessary to use the one giving maximum movement, in this country. The position which minimizes the stroke of the pump should be re-used for summer.

Engines give best efficiency on cold air; but in damp, frosty weather this can lead to carburettor icing and misfiring. So, what is called a hot-spot is provided on many engines, to feed air from a hot region of the exhaust manifold in winter. Various methods are used – sometimes the air filter has to be undone and re-positioned so that the intake is near the exhaust for winter; in others a more convenient summer and winter change-over lever is provided in the air intake to the filter. Consult the handbook, and make the necessary adjustment for summer or winter.

The crankcase ventilation valve

Combustion fumes tend to find their way past the pistons into the crankcase, and on older engines a pipe will be found at the side of the engine, which leads these fumes out below the car. With very worn engines running at speed on a motorway, clouds of blue smoke may be seen emerging apparently from the underneath of the car; they are coming from the crankcase breather.

In the interests of reducing pollution, it has been decreed that these unburnt oil-laden gases will not just be evacuated into the atmosphere. Instead, they are to be fed back into the engine to be burned off. A valve, usually made by Smiths or A.C.-Delco, is interposed between the crankcase and the feed to the carburettor intake, to prevent too much air from being admitted at low speeds or at tick-over. It allows maximum flow when the accelerator is hard down, when maximum ventilation is needed.

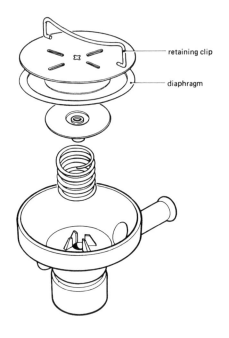

retaining clip

diaphragm

Progressive accumulation of oil and gummy deposits may eventually prevent the valve from closing properly, and extended neglect of the valve is often the cause of inexplicably bad tick-over in some poorly-serviced cars. Hoses and connections of the valve should be disconnected, and on removal it is easily taken apart on the bench and cleaned-up using cloth moistened with meths. Avoid getting spirit on to the diaphragm, and check that the diaphragm is undamaged. When re-assembling the Smiths valve, which is held together by a spring clip, make sure that the needle, like that of an 'S.U.' carburettor, engages with the orifice correctly, and that it is not damaged.

Tappet adjustment

To allow for the expansion of metal as it heats up, a gap or clearance has to be allowed in the engine between the valve stem and its operating tappet. This is called the valve clearance, and a periodical check has to be made to ensure that the gap is correct. The usual tendency is for the clearances to reduce, which can lead ultimately to valves not closing properly and valve burning as a result. If the gaps widen, the tappets become noisy, spoiling the refinement of the car.

Basic handbook information for the car usually includes the recommended clearance, often different for inlet and exhaust, and states whether it should be checked hot (as with most Ford

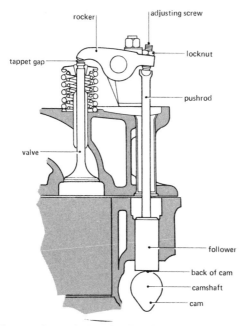

Tappet layout of a typical pushrod engine.

107

engines) or cold. An extended cooling-off period must be allowed – preferably overnight – before cold tappet adjustment may be made, while hot adjustment means at or near normal running temperature. With pushrod-operated valves as on the majority of cars, adjustment is made very easily by inserting the appropriate thickness of feeler gauge between the rocker and the end of the valve stem, slackening the lock-nut at the other end, adjusting with the screw, and then re-tightening the lock-nut.

More and more cars now have overhead camshaft engines, which give better efficiency but make adjustment more diffi-

Many pushrod ohv engines have simple screw and locknut tappet adjusters. Each tappet must be adjusted when the appropriate corresponding valve is fully open (see text for table). Slight drag on the feeler gauge indicates a correctly set gap, which should be rechecked after tightening. This engine is in a Morris Marina.

cult. Exceptions are the overhead camshaft Vauxhall engine which is simply adjusted by turning a screw within the camshaft follower, and the BMW system of using an eccentric follower in the end of the rocker which is turned for adjustment. On others, where the rocker is, in effect, a finger beneath the camshaft, adjustment is made quite easily at the pivot end. The difficult adjusters are those using biscuit shims – pieces of thin metal of different thicknesses between the camshaft and the inverted valve buckets. The elaborate procedure of measuring the existing gaps and making a note of the necessary adjustment of thickness, then removing the camshaft and fitting the appropriate new shims, has to be followed.

Some engines such as this Vauxhall Viva have a simplified means of tappet adjustment, done by turning a self-locking nut on the central pivot of the rocker. Again, each valve must be adjusted with the relevant corresponding valve fully down.

Vauxhall Viva rockers have no provision for adjustment at the valve; instead, the whole rocker is adjusted up or down by raising or lowering the central dome-shaped nut against which the rocker pivots. This is quite easily done from above, by rotating to raise or lower the self-locking nut which holds the pivot down. The same arrangement is used on the Ford V4 and V6 engines.

For many years Vauxhall six-cylinder engines had to have their tappets adjusted with the engine ticking over, which was difficult as the intermittent opening and closing of the valve being worked on tended to jerk the screwdriver and spanner; and the feeler gauge used to be damaged in the process. For most current engines, however, adjustment is done with the engine at rest, and it is essential that each tappet is adjusted while the follower is exactly on the back of the cam, and this is determined by the position of a corresponding tappet.

Once the rocker cover has been removed the valve stems and springs will be revealed, and a glance at the ports of the inlet manifold and the exhaust manifold (usually rusty because the heat burns off any protective coating) reveals the arrangement of valves. The question to ask is: does the engine have the inlet valves in pairs, with an exhaust port at each end of the engine? Or, do inlet and exhaust valves alternate, so that at one end of the engine there is an inlet valve and at the other an exhaust? The question is important because it determines which valve must be fully open (tappet right down) when adjusting a corresponding valve.

Having determined to which category the engine belongs, as listed opposite, the appropriate column in the table on p. 112 can be consulted and the engine turned to open the correct corresponding valve for adjustment of each valve in turn. For example, if a four-cylinder in-line engine has an exhaust valve at each end and the inlet valves in pairs, it is in Column A in the table. We number the valves from the front (or pulley) end of the engine. With 'V'-engines such as the V4 and V6 units used by Ford, the valves are numbered from the front on the right-hand bank of cylinders first, as seen if you were standing behind the engine. To avoid confusion, lean over with your head at the back of the engine looking towards the front of the car. The cylinders now to your right are called the right bank. After number four valve (or number six in the case of a V6), we continue numbering from the front, now dealing with the left bank.

It will be seen from the table that the order for corresponding valves is the same for a V4 (Column E) as it is for a four-cylinder in-line engine with paired inlet valves and an exhaust at each end (Column A). Using these as examples, turn the engine manually using one of the methods advised in Distributor Maintenance, until valve number eight is fully down; you are now ready to adjust tappet number one. The sequence continues as given in the table, for which the following is the key.

A =Four-cylinder engines with an exhaust valve at each end, i.e. 1 ex – 2 in – 3 in – 4 ex – 5 ex – 6 in – 7 in – 8 ex.

B =Six-cylinder engines with an exhaust valve at each end, i.e. same as A and continuing 9 ex – 10 in – 11 in – 12 ex.

C =Four-cylinder engines with alternate inlet and exhaust valves, i.e. 1 ex – 2 in – 3 ex – 4 in – 5 ex – 6 in – 7 ex – 8 in.

D =Six-cylinder engines with alternate inlet and exhaust valves, i.e. same as C and continuing 9 ex – 10 in – 11 ex – 12 in.

E =V4 engines, beginning with the right bank.

F =V6 engines, beginning with the right bank.

When adjusting this tappet	This tappet must be fully down (valve open) Use appropriate column for type of engine					
	A	B	C	D	E	F
1	8	12	7	11	8	10
2	7	11	8	12	7	9
3	6	10	5	9	6	12
4	5	9	6	10	5	11
5	4	8	3	7	4	8
6	3	7	4	8	3	7
7	2	6	1	5	2	6
8	1	5	2	6	1	5
9	–	4	–	3	–	2
10	–	3	–	4	–	1
11	–	2	–	1	–	4
12	–	1	–	2	–	3

Naturally there are only eight tappets in a four-cylinder engine (columns A, C, and E) except in the rare cases of cars such as the Triumph Dolomite Sprint which have twin inlet and exhaust valves to total sixteen. Some engines have no need for tappet adjustment at all, because a self-compensating hydraulic system is used. This applies to most V8 engines such as the Rover and Rolls-Royce, and to a number of in-line engines, as in the Opel Rekord and Commodore range.

After tappet adjustment has been completed it is a wise precaution to make a final check that all lock-nuts are tight, and then quickly run through them all with the feeler gauge again to check that the sequence was right. One of the tappets may have been adjusted with the wrong fellow open, and this last check will reveal it.

Top overhaul

In the days of poor-quality petrols and oils, decarbonizing used to be a fairly regular service procedure for an engine; but improved technology and advanced engine design have practically eliminated this requirement. Mileages of 60,000 to 80,000 without disturbing the cylinder head are commonplace in high-mileage use, and usually when a top overhaul is finally needed, attention to pistons, rings, bores and bearings may be due as well. Occasional need to carry out a top overhaul may result, however, from the burning of an exhaust valve.

Be quite sure first that the trouble has been correctly diagnosed – this usually involves connecting a compression test gauge to the sparking plug hole after removal of the plug, turning the engine several times on the starter, and comparing readings for all cylinders. Readings are given in psi (as for tyre pressures), and such results as: 126 – 122 – 124 – 94 would certainly confirm loss of compression in number four cylinder, for which cylinder-head removal would be the next stage in analysis, giving access to the valves if these are at fault.

Removing the cylinder head of an engine having push rod-operated valves is fairly straightforward work within scope of the home mechanic who has reasonable skill and the appropriate tools. In particular, a torque wrench is required, so that when re-assembling, the cylinder-head nuts can be turned to the correct degree of tightness; and a valve spring compressing tool will be needed to allow each valve spring to be compressed slightly in turn, while the retaining clips or cotters are released. A suction-fitting valve-grinding tool – little more than a round piece of wood with a sucker at the end – is needed to enable the valve to be revolved by sliding the hands to and fro to grind the valves with a little valve grinding paste, against their seats. This makes the perfectly bedded joint necessary for effective sealing of the valves.

Details of the manufacturer's recommended procedure are helpful so that no important stage is omitted. For example, one might find that an oil feed pipe to the rocker gear has been left connected if the stage-by-stage order of dismantling and re-assembly is not followed.

On some push rod-operated engines a cam follower can be disturbed by careless removal of the push rods, resulting in unnecessary further dismantling work to enable the follower to be set back on the cam. Take care when lifting the rocker assembly that the push rods remain in place; then lift each push rod about $\frac{1}{2}$-inch and give it a smart tap downwards; this should knock the follower free and ensure that it is not pulled out of its housing as the push rod is withdrawn. Many overhead camshaft engines have provision for withdrawing the camshaft without disturbing the timing chain, and the work is again within scope of the reasonably knowledgeable amateur; but as advised before, be sure that you know the recommended procedure and have the appropriate tools before embarking upon the work.

4 · Transmitting the Power

POWER is transmitted from the engine to the road wheels in a car by means of three main components, all subject to stress and wear. The clutch allows the smooth mating of the rotating engine and the stationary car; the clutch also provides the means of temporarily interrupting drive from the engine. The gearbox provides the long-term interruption of engine power with its neutral position, and it offers the choice of lower gears necessary to multiply the engine's effort when needed for acceleration or hill climbing. Included with the gearbox may be an overdrive, which is simply an alternative gear with conveniently handy electric switch control. Both clutch and gearbox may be replaced by a hydraulic torque converter, which gives variable drive according to engine speed, from slight creep at tick-over to full power at about 2500 rpm, coupled to a more advanced gearbox having a controlling brain and automatic selection of gears according to speed, load and accelerator position. This, of course, is the automatic transmission, seen at its simplest in the Daf, which has rubber belts running between pulleys of variable size.

The third component is the final drive, comprising the differential, which allows the power to be transmitted to one wheel on each side of the car while each is turning at different speeds while going round a corner; and the connecting links or shafts which carry the power from the output shaft of the gearbox to the wheels. In cars like the Mini and Imp, where the engine is near to the driven wheels (front and rear respec-

115

tively), no long propeller shaft is needed; but there must still be a shaft from the differential to the driven wheel on each side. These in turn must be articulated, to allow the wheels to rise and fall over unevenness in the road surface without bending or breaking the shafts. In the most common layout, of course, the shafts from differential to wheel are enclosed in a rigid unit, called a 'live' back axle. The word 'live' tells us not – as many people think – that it jumps about over the road surface in a lively fashion, but that it transmits the power. If it does not transmit power, as in the front-wheel drive Audi, it is called a 'dead' back axle.

Though exposed to road dirt and spray, and frequent reversals of thrust – one minute the engine propelling the car, the next the momentum of the car turning the engine, with the driver's foot off the accelerator – these components give generally good service and require surprisingly little in the way of routine maintenance. Some attention, however, is needed and must not be overlooked. Because a lot of dismantling may be necessary to obtain access to transmission components, overhaul when needed is likely to be expensive.

Clutch operation

Between the flywheel of the engine and the plate which transmits the engine power to the input shaft of the gearbox, is the clutch plate, lined on both sides with friction material not unlike the material used for drum brake linings explained earlier; but the lining material in this case is distributed all round the circumference, and on both sides. In technical terms, the clutch plate is often called the centre plate, for the obvious reason that it lies between the flywheel and the outer plate; more correctly, it is called the driven plate. It is easy to imagine the three plates in contact, and hence all turning at the same speed; this is the normal running condition. They are held in contact by powerful springs all the way round the outer (or pressure) plate, and there is direct drive from the engine to the gearbox.

It takes a little more thought to visualize what happens when the driver presses his foot on the clutch pedal. Hydraulic and/or mechanical linkages are used to move the pressure plate (the one farthest from the engine), away from the centre plate. As they move apart, they become free to turn at different speeds; the drive from the engine to the gearbox is now interrupted, and equally obviously it can be seen that the drive is taken-up again, smoothly, as the driver gently releases his pressure on the clutch pedal.

Two important points emerge. First, as soon as light pressure is applied to the clutch pedal – and long before the clutch begins to disengage – the clutch operating fork begins to

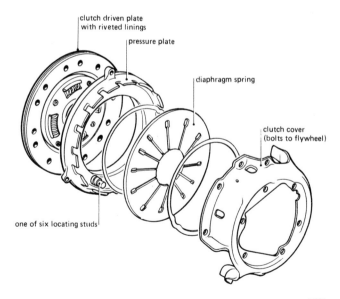

clutch driven plate
with riveted linings

pressure plate

diaphragm spring

clutch cover
(bolts to flywheel)

one of six locating studs

117

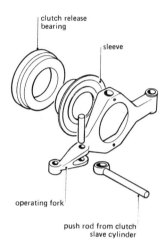

clutch release
bearing

sleeve

operating fork

push rod from clutch
slave cylinder

A diaphragm spring clutch ball race release bearing.

move what is called the clutch release bearing, bringing it into
contact with the clutch release levers. Because it is a ball bearing
(on most cars these days) it can cope with the fact that the operating fork which holds it is at rest, although the clutch release
levers are whirling around at the same speed as the engine
flywheel. But it is easy to see why it is important not to rest
your left foot on the clutch pedal while driving. If you do, the
free play is taken up, the release fork is in contact with the
bearing, and the bearing is turning all the time, all resulting
in unnecessary wear.

The second point is that in the partially-released position,
as when taking up the drive to move away from rest, the
linings of the clutch centre plate are slipping and wearing.
They are only in a non-wear condition when the clutch pedal
is either fully released or fully depressed. So it can be seen
also why it is wrong to slip the clutch any longer than neces-

118

sary for smooth drive take-up, and why holding the car on the clutch instead of the brake on a slope, for example, causes unnecessary wear.

Most cars nowadays have what is called a diaphragm spring to hold the plates firmly together, instead of the former arrangement of a series of coil springs. The force exerted by the diaphragm spring extends over a wider range of movement, helping to make the clutch pedal lighter to work, and compensating for wear to a greater extent than was possible with the coil springs.

Over the years, all kinds of means have been used to transmit the movement of the clutch pedal to movement of the clutch operating fork; cables, rods, hydraulics, and even chains have been employed. Cables are becoming popular again, and the handbook should be consulted to see whether any lubrication is needed. Where hydraulic connection is used, a second hydraulic cylinder will be seen, under the bonnet, usually alongside that for the brakes. The same rules as for hydraulic brakes apply. In summary, remember to check the level in the cylinder, though as safety is not involved a less-frequent check may be made. Ensure cleanliness, and only top-up with hydraulic fluid of the recommended grade for the car. Bleeding is carried out when needed, as for brakes, and the bleed valve will be found usually at the back of the slave cylinder (the cylinder mounted near the clutch unit, connected to the operating fork).

As with brakes, a sudden marked drop in the fluid level is a warning that leakage has developed; just topping up is not a cure, and the problem will almost certainly get worse unless something is done. The most likely cause is failure of the seals in the slave cylinder, whose piston necessarily travels over a much bigger range than the slave cylinders in the brakes. Follow the pipe from the master cylinder, looking for signs of leakage, all the way to the slave cylinder. Watch the slave cylinder while an assistant operates the clutch pedal with the engine at rest, and look for escaping fluid. Replacement seals will be needed if this is the case, but this is more in the category of a repair than of maintenance, and a job for a mechanic. After new seals have been fitted, bleeding is necessary to remove air from the system. Bleeding is also needed if the level has inadvertently been allowed to fall so low that air is pumped into the system instead of fluid.

Occasionally an expensive clutch failure is diagnosed when all that is involved is a fault in the operating system. In anticipation, it is a good idea some time to have an assistant operate the clutch pedal while you watch the movement of the clutch-operating fork, underneath the car. The amount of travel while

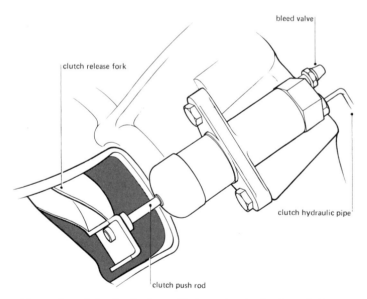

Connection of the clutch slave cylinder to the clutch release fork.

the system is operating correctly is then known, and the difference if a fault occurs later will be very apparent. An indication of the trouble in a clutch is given by the abruptness, or otherwise, with which it develops. Sudden failure, developing in the course of a journey or a couple of days' use, is almost certainly a failure in the operating system.

In addition to a hydraulic leakage, failure can result from breakage of the piston return spring in the master cylinder. Again, renewal, and fitting of new seals, followed by bleeding, will be necessary. If a cable is used to operate the clutch, breakage of the cable can be a reason for clutch failure, though as manufacturing techniques improve, this becomes less likely.

Clutch adjustment

It is essential that when the clutch pedal is released, the full clamping load of the springs in the clutch unit are able to bear against the centre plate; otherwise unwanted slip may occur and the rate of wear will be needlessly increased. To ensure that this is the case, manufacturers usually specify that there should be at least an inch of free play at the clutch pedal, when a mechanical linkage (cable or rod), is used. When a hydraulic system is used, the free play is less easily checked, as there is usually no wasted movement before the piston begins to move in the master cylinder. Free play has to be checked at the slave cylinder. Clearance between the end of the clutch-operating lever and the end of the slave cylinder push rod should not be more than about 0·10 of an inch – very different from the 1-inch clearance when measured at the pedal.

No adjustment at all is provided on some cars, but on others the adjustment may be by means of slackening a clamp which holds the slave cylinder, and re-positioning the slave cylinder

121

to provide the correct clearance; on others it is done by slackening lock-nuts on the push rod itself and adjusting the length of the push rod. In this case, the clearance is obtained at the other end of the push rod, where it abuts against the piston in the slave cylinder; the rubber boot over the end of the slave cylinder will have to be removed for the end of the rod to be seen.

As the lining material wears (albeit slowly, over thousands of miles of normal driving), the pressure plate when in contact with the centre plate will move nearer to the flywheel. Because of the pivoting action of the clutch release levers, this will tend to diminish the amount of free play present in the operating system, so when checked, this indicates the need for adjustment. If no provision is made for adjusting the clutch, it is presumed that the system provides sufficient travel for the amount of wear likely to be encountered before renewal of the linings becomes necessary, or is self-adjusting. The handbook will usually indicate whether provision for adjustment is made or not.

Clutch lining renewal

One of the most frequent causes of rather costly work on the modern car is the need to renew the lining material on the clutch centre plate. On most cars this involves removing either the engine or the gearbox. Only on some front-wheel drive cars which have the gearbox in unit with the engine, is it sometimes possible to obtain access to the clutch plate without major dismantling, and even on some of these the engine has to be removed.

On cars like the Sunbeam Talbot and Triumph Herald or Vitesse, it was possible to remove the gearbox cover and obtain access to these units from above; but the more usual arrangement is that it has to be attacked from below. The

122

clutch bell housing is unbolted from the engine, the gearbox output shaft is disconnected from the propeller shaft, and the complete gearbox and clutch bell housing unit is withdrawn, giving access to the clutch, which remains bolted to the engine. The work is within scope of the more competent home mechanic, since few special tools are required, but a reasonable knowledge of the car or a study of the workshop manual is invaluable.

It is beyond the scope of this book to cover the detailed procedure of stripping down to replace the clutch plate, since specific directions for the particular car are needed, and may differ one to another. But for those who tackle the work some precautionary pointers may help.

It is usually possible to remove the gearbox without removing the propeller shaft. The self-locking nuts at the front coupling are undone, and a chalk mark should be made across the flanges so that reconnection can be made with the faces mating in the same position. This applies also to the clutch assembly – chalk a line across to mark its position in relation to the flywheel – otherwise the flywheel balance may be upset. When re-assembling, new bolts and self-locking nuts must be used for the propeller shaft joint.

If the instructions advise taking the weight of the rear part of the engine on a jack, make sure it cannot slip, and that a strong piece of wood is interposed to avoid damaging the sump. Check that it is supporting the weight of the unit adequately before undoing the rear mounting. When altering the vertical height of the rear of the engine, take care that the fan is not brought into contact with the radiator. Both when removing and re-fitting the gearbox, do not let it hang suspended by the input shaft; it must be supported until the locating bolts have been fastened.

If the gearbox is removed just for attention to this unit or the selectors, for example, the clutch may be left undisturbed.

Provided the clutch pedal is not pressed while the gearbox is out, it should be possible to push the gearbox input shaft back through the clutch plate to engage with the flywheel. In most cases, however, when the gearbox is removed the opportunity will be taken to inspect the clutch plate, even if this was not the main reason for dismantling. As soon as the clutch cover is unbolted, the spring pressure against the plate will be released, and as the gearbox has been removed and the shaft is no longer there to hold the clutch plate in alignment, the plate will literally fall out.

The gearbox can never be replaced until the clutch plate is in exactly the right position, so a clutch-aligning tool is needed to take the place of the gearbox input shaft. This is simply a splined shaft without the gearbox attached; it is inserted with the clutch plate already on it, and engaged with the hole in the centre of the flywheel. The clutch cover then can be re-fastened, clamping the clutch centre plate into the right position. The aligning tool can be withdrawn because the plate is now located by the pressure plate, and the gearbox can be offered-up.

This explanation of the need for an aligning tool necessarily

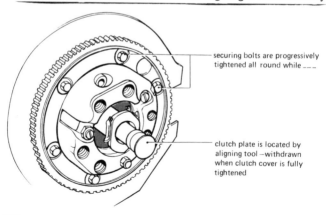

securing bolts are progressively tightened all round while ___

clutch plate is located by aligning tool —withdrawn when clutch cover is fully tightened

preceded the advice on removing the clutch cover, in which the important thing to remember is that the bolts should be slackened progressively, a little at a time, all the way round. As they are slackened, the enormous spring pressure exerted against the centre plate is released, and hence the reason for undoing the cover progressively. Similarly, when re-fastening, the tightening-up must be done progressively.

Unless a clutch plate proves to be in little-worn condition, renewal is usually good economy since a new one (there is usually no exchange, as there is for brake shoes) may cost only about £10 to £20 a small sum in relation to the effort or cost of obtaining access to it. At the same time, the thrust race should be inspected and renewed if its condition is at all suspect.

Clutch judder

Jerky clutch take-up is a tiresome fault, which causes undue stress on the engine and other transmission components. First, one should identify whether the problem really *is* judder in the clutch, or whether it is engine shake; the two are not easily distinguished. Engine shake is visible from outside, if you do some practice starts with the bonnet open while an observer stands alongside. Causes may be a loose engine tie rod, or perished and broken engine mountings. When it has been determined that the problem really is clutch judder, check for smooth action of the clutch release mechanism by observing underneath (car secure and engine switched off) while an assistant works the clutch pedal. Lack of lubrication in a mechanical clutch linkage could be the cause.

If it is established that there are no visible causes, inspection is advisable, involving the removal of gearbox (or in some cases the engine), as previously explained. Careless assembly can result in a buckled clutch centre plate; this, worn linings, or oil on the linings, can cause judder.

125

Clutch life

Because you cannot easily inspect the clutch plate to see how worn the linings are, as you can with drum brakes, it is not easy to know when attention is needed. The best guide is that if the clutch is taking up smoothly, and there is no tendency to slip once the pedal has been fully released, it may be presumed that all is well. The free travel mentioned under Clutch Operation is also a guide, and when all free travel has gone or available adjustment has been used up, it may be guessed that the linings are getting pretty thin. Clutch judder, explained in the previous section, or excessive clutch slip are other signs that the extent of wear requires attention.

Once clutch trouble of this kind is experienced, and the clutch-operating mechanism has been checked and found in order, inspection should not be too long delayed. Extended use of a worn-out clutch can eventually result in costly damage to the smooth faces of the flywheel and pressure plate.

It is impossible to generalize about the life of a clutch, as so much depends on the conditions in which the car is used, and the competence of the driver. Abuse or extensive traffic work may wear out a clutch in 15,000 miles or less, while some drivers happily make a clutch last 60,000 or 70,000 miles. Within these extremes, 25,000 to 35,000 miles may perhaps be described as a norm for clutch life.

Gearbox lubrication

Some cars like the British Leyland Mini and 1100 have the gearbox in unit with the engine, and one lubricant serves for everything. The more usual arrangement, however, is for the gearbox to be separate, incorporating its own special lubricant. Periodic renewal of the gearbox oil used to be standard service procedure on all cars, but with improvements in oil tech-

126

nology the tendency has been to eliminate this and say that the oil is good for the life of the gearbox. The handbook or lubrication chart should be checked to see whether provision is made for changing the oil, and if so what grade is recommended. Some gearboxes take engine oil (SAE 30 grade, for example), while others have much thicker gearbox oil – such as SAE 90.

Contents will usually be about $1\frac{1}{2}$ pints, but again the amount differs from one model to another. However, it will certainly be found that the Castrol plastic dispensers containing oil of the correct grade and supplied complete with a transparent plastic spout, make the job of re-filling, which generally has to be done from underneath, much easier.

Most gearboxes are provided with two tapered, square-headed plugs. The square head identifies them clearly from other hexagonal nuts which form part of the gearbox fastenings. One of the plugs is located at the bottom, either underneath the casing or at the side; the other is positioned higher-up, and serves as combined level and filler plug. The area around this upper plug should be cleaned first, to prevent dirt from falling into the gearbox; then unscrew the plug. As well as speeding the draining-off process, removal of the filler plug before the drain plug ensures that one is not left with no oil in the gearbox if, for any reason, the filler plug cannot be removed.

A tray or oil drain reservoir is then placed underneath the gearbox, and the drain plug removed. A ring spanner of correct size will be needed to start turning it, and then it can be unscrewed the rest of the way by hand – a messy job, as oil begins to leak past before the nut is completely clear. The easiest way is to screw it out so far, then make the final release with the spanner, and not worry unduly about the plug falling into the oil tray. It can be fished out later, but of course it is vital not to throw it away with the old oil by mistake!

127

Waste oil must be disposed of properly, not poured down drains. One way is to bury it in your garden, and let it seep away, although it is preferable to hand it in to one of the many garages which now accept waste oil. Draining-off is best done when the gearbox is warm after a run; but this is not always practicable, in which case at least an hour should be allowed for the final drips of contaminated oil to run out. The drain plug is then replaced and tightened, and the new oil is squeezed in from the dispenser bottle through the filler hole. When approximately the right quantity of oil has been delivered the cut-off point will be identified by sudden over-flowing of oil running down the side of the casing. The filler plug is now replaced and tightened as well. Points to note are that the tapered plugs should not be over-tightened – screw them in firmly with a short spanner, but *not* as though your life depended on it; and that re-filling should be done with the car on the level, not at an angle on wheel ramps.

Drive-line lubrication

Even after all steering and suspension grease points had been eliminated, many car manufacturers still considered it neces-sary to provide for periodic greasing of the sliding splines on the propeller shaft. Again, the handbook or lubrication chart should be checked to see if this applies to your car, and if so the grease point should be given three or four strokes with the gun as recommended. If the car is on level ground it is permis-sible to locate a wheel chock or brick about a foot ahead of the front wheel and behind the rear wheel, release the brake, then get underneath and gently push or pull the car as required to bring the propeller shaft grease point into an accessible position. In most rear-wheel drive cars, the rear hubs are lubricated by the back axle oil, but some with independent rear suspension, like the Triumph Herald family have separate

128

grease points for the rear wheel hub bearings. The handbook
should be studied, and it may be found that a blanking plug is
fitted; if so, the area around it must be cleaned, then the plug is
removed and a grease nipple is screwed in. Two or three
strokes of the gun are given – excess grease is to be avoided for
fear of contaminating the brake linings – and the blanking
plug replaced.

Back axles

Periodic changing of back axle oil has generally been aban-
doned due to improved longer-life oils, and incidentally
because it was found that more harm was done by allowing

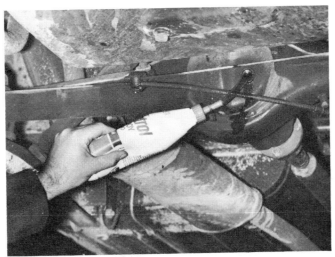

*Use of a Castrol squeezy filler bottle (oil dispenser) with flexible
spout simplifies the often difficult task of topping up a final drive
unit or gearbox from underneath. Provision of a common filler and
level plug, and omission of the drain plug, are common practice.*

129

dirt to fall in when changing the oil than by leaving the old oil in. It may be found that a drain plug is fitted and provision made for regular oil changing, in which case the procedure followed is much as for the gearbox, explained earlier.

A more usual arrangement is for a single plug to be fitted approximately halfway up the domed cover of the differential unit or its casing, which can be removed to allow a check of the lubricant level. Again, the area around the plug must be cleaned before the plug itself is removed, and special care should be taken to prevent entry of any dirt dislodged from the area above it. A clean screwdriver or any other straight implement too long for there to be any risk that it could fall inside is then used as a dipstick. Alternatively new oil can be squeezed in from the dispenser, and any excess will flow out of the orifice.

As well as the annual check on back-axle oil-level, it should also be inspected if signs of oil leakage from the front seal (adjacent to the propeller shaft) or from the region of the wheels is seen. Similarly if there is a marked increase in back axle whine.

Many cars such as most Fords, for example, have no filler level plug on the back axle: the initial oil is intended to last the life of the unit, and renewing the oil is only possible as part of re-assembly procedure.

Gearbox renewal

It is considered that only a knowledgeable amateur or professional mechanic would consider gearbox *overhaul* within his capabilities, especially as special tools are often necessary to press new bearings into position; but gearbox *replacement* with a reconditioned unit may well be within scope of the enthusiast. The same points already covered in Clutch Lining Renewal should be observed, and in view of the substantial work

130

involved it makes sound sense for the clutch plate and release bearing to be inspected at the same time. To avoid messy spillage, as well as to help by slightly reducing the weight, it is recommended that the oil be drained out first, and the drain plug then re-fitted.

Once again, it is emphasized that unless the person doing the work knows what he is about, he should study a workshop manual first, to find out the exact order and procedure to be followed. Otherwise it is easy to get involved in unnecessary labour, such as removing the propeller shaft or engine when the gearbox can in fact be removed, on a particular model, with one or both of these items *in situ*. Similarly, details such as disconnection of the speedometer drive cable may get overlooked. Before the gearbox can be removed from the car, a preliminary requirement is disconnection of the gear-change mechanism, and here again the manufacturer's specific instructions are invaluable.

Overdrive

Production of the Borg Warner overdrive ceased a long time ago, and current units are all of the Laycock-de Normanville type, with electric solenoid operation. The overdrive contains oil as well as the gearbox, but the lubricant is shared and no separate draining and oil-changing procedure is necessary. However, it is usual for the overdrive to have its own drain plug, which must be undone as well as the gearbox drain plug, when changing the common oil; otherwise, that part of the oil which is in the overdrive will be trapped, and will mix with the fresh oil. This drain plug may be hexagonal, and care should be taken to ensure that the correct plug is undone, as there may be other nuts nearby which are not to be disturbed. Look for the word DRAIN adjacent to the plug, and consult the handbook.

131

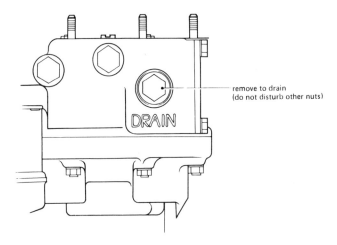

remove to drain
(do not disturb other nuts)

DRAIN

*Where an overdrive drain plug is provided it may be identified
by lettering on the casing.*

No separate filler plug for the overdrive is provided, but
after filling-up the gearbox it is advisable to take the car on a
short run, engage overdrive a few times, and on return re-check
the level. It may well be found that more oil can now be
squeezed in, before it begins to run back out of the level filler
orifice.

No routine maintenance for an overdrive is necessary, and in
general these units give excellent service. If trouble is
experienced it may be nothing more than need for adjustment
of the operating solenoid. Examine the overdrive unit, situated
to the rear of the gearbox, and look (on the right-hand side)
for a cylindrical unit not unlike the clutch slave cylinder of a
hydraulically-operated clutch, but fed by an electric wire
instead of a hydraulic pipe. Just ahead of it will be seen a small
inspection plate, which can be removed to reveal the operating
lever and its connection with the solenoid plunger.

Halfway along this lever will be seen a swaged hole which

132

may look as though it should have a bolt through it. This is, in fact, an aligning hole to enable the lever to be correctly positioned. Behind it is a matching hole (or indentation) in the casing, and when the overdrive is engaged the two should be in line. Check first that current is getting to the solenoid, by selecting top gear (car at rest and engine off), switching on the ignition, and working the overdrive switch. A sharp click should be heard each time the overdrive is switched on. If not, and a check underneath shows that the lever is not moving, check the wiring through with a test probe to locate the fault.

If the solenoid is operating correctly, leave it now in the switched on position, and check underneath the car that the hole in the lever is in line with the one in the casing. A short piece of rod of 3/16 inch diameter can be used to register the two holes, and the adjusting nut is now screwed back so that –

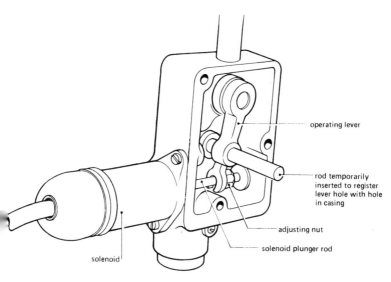

operating lever

rod temporarily
inserted to register
lever hole with hole
in casing

adjusting nut

solenoid plunger rod

solenoid

Adjusting the overdrive solenoid.

133

still with overdrive theoretically engaged – the nut just contacts the operating lever. This ensures correct positioning of the lever at the full travel of the solenoid. Wrong adjustment of the solenoid can result in blowing the fuse, as it is operated by a very heavy current, switching automatically to a small holding current at full travel. If incorrect adjustment prevents the solenoid plunger from reaching full travel, the big operating current remains switched in. After adjustments have been completed, of course, the aligning rod is removed and the inspection cover replaced.

Automatic transmission

Replacing the clutch and manually-shifted gearbox on increasing numbers of cars is automatic transmission, of which it is a fair generalization to say that it is less troublesome and requires less maintenance. By eliminating the friction of engaging and disengaging a clutch, the wear rate is reduced, and unless the driver of a manual transmission car is particularly skilful, the automatic transmission is kinder to the rest of the machinery. Because of the complication of the automatic gearbox and the need for special tools when dismantling it, repairs are likely to be beyond the ability of the owner-mechanic, and to be decidedly expensive. This is the debit side of the picture, which makes it all the more important for the owner to seek competent professional advice if he notices any marked change in behaviour or additional noise emanating from the automatic transmission. For example, it may suddenly begin to slip and surge between gear changes when accelerating hard, indicating that the brake bands, which enable power-sustained gear changes to be made, are slipping. Caught early the problem may be cured by simple external adjustment; but ignored, it may lead to accelerated wear and ultimate costly overhaul.

A check of the handbook will usually reveal that no routine changing of the transmission fluid – don't call it oil – is needed. It is important, however, to carry out a periodical check of the fluid level, for which a dipstick is usually accessible under the bonnet, at the back. Again, refer to the handbook to see how the check is to be made. Invariably it specifies that the fluid should be warm, i.e. after a few miles' run, and that the engine should be running. Some specify that DRIVE should be selected, in which case an assistant must be sitting in the driving seat. Others are to be checked with PARK selected. There are usually two marks on the dipstick, and after wiping with a clean cloth and inserting fully, the fluid mark should lie between them.

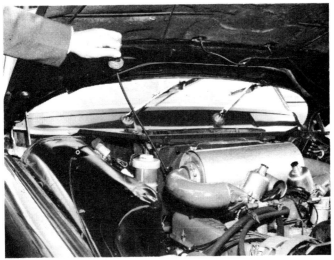

Cars with automatic transmission invariably have a separate dipstick for the transmission fluid. Handbook instructions should be followed, as the check is usually made with the engine running and may require Drive or Neutral to be selected. A clean non-fluff cloth should be used to wipe the transmission dipstick.

135

If it is down near the lower mark, topping-up is needed, when two points should be noted: first, ATF (Automatic Transmission Fluid) of the recommended make and grade must be used and never oil, hence the comment made earlier; second, that any containers or funnels used for pouring the fluid must be scrupulously cleaned. Re-check with the dipstick after adding about a $\frac{1}{2}$ pint using the same procedure as before, and avoid over-filling.

Need for frequent topping-up tells of a leak somewhere, which should be investigated before it gets worse and all the fluid is pumped out as you drive along. If a remote cooler is fitted for the transmission fluid, check periodically the condition of all flexible pipes, and look out for any signs of leakage or chafing. Underneath the body of the automatic transmission will be seen some gauze-covered vents, which allow waste heat to escape from the torque converter; an occasional check should be made, when under the car, to see that these do not become blocked by mud or engine oil.

Apart from these basic attentions, automatic transmission is for the greater part maintenance-free.

Universal joints

It is outside the scope of this book to cover the removal and stripping-down of universal joints, but as the home mechanic must also be his own inspector, a periodic check of all joints and couplings in the propeller shaft and rear half shaft joints (if the car has independent rear suspension), or front-drive shafts (if it has front-wheel drive), should not be overlooked. Check with a ring spanner that the flange nuts on the propeller shaft coupling keep tight; if they come loose, new self-locking nuts should be fitted and the bolts should be replaced at the same time.

Do not be alarmed at the presence of quite a bit of free

movement when you attempt to turn the propeller shaft of a car from underneath, with the brake firmly on; tnis is normal backlash in the final drive and differential gears. But there should be no looseness evident, or clicking noises, when one side of a universal joint is held while you try to turn the other side. A worn universal joint may eventually make itself evident in the form of transmission vibration at certain speeds. Replacement will usually be a job for a professional mechanic. These joints usually have their own lubricant sealed in, and lubrication is needed only to the sliding splines mentioned earlier – but again, check handbook or lubrication chart.

With a front-wheel drive car, check from time to time the condition of the flexible gaiters at the outer ends of the drive shafts; if they are seen to crack or split and allow grease to escape they must be replaced as soon as possible, or the joint itself will wear rapidly. When driving a front-wheel drive car, be on the alert for clicking noises when the car is turning on full lock. This tells of worn front drive joints, which should be replaced without delay; and again this is a job for the professional.

5 · For the Sake of Appearances

WHEN the average used-car buyer surveys a prospective purchase, he invariably checks the mileometer reading and presumes that the mechanical condition will be average for the mileage covered. A trial run, if made at all, will be simply to check that no major mechanical faults are evident, and even in many cases for the buyer to confirm for himself that he likes the particular make and model chosen as well as he thought he would. It is almost entirely on bodywork, therefore, that the condition of the car and its estimated value will be assessed. Equally obviously, it is to the bodywork that particular care in maintenance must be given if one wants the value of the car to be kept up to standard. Neglect soon leads to rapid deterioration, which in turn means erosion of the car's true value. Depreciation is rapid enough, without adding to the rate of drop in value by letting a car look shabby and older than it really is.

Paintwork

Big improvements have been made in the standards of paint finish for popular cars from the point of view of durability; materials are better and so is the basic preparation of the steel body prior to painting. But this does not mean that no care is needed for the optimum life to be obtained from the initial paintwork. Paint is essentially porous, and if not protected will eventually allow damp to find its way through to the metal

138

beneath, resulting in rust formation beneath the paint, ultimately to break through the surface as an ugly blister.

The first essential is removal of dirt by washing at reasonably regular intervals. If left to accumulate, mud holds moisture and accelerates corrosion. Actual harm can be done, however by the simple process of washing. The bucket of grimy water and a grit-laden leather can easily cause more surface scratching than exposure to one of the more destructive of the automatic car-wash machines. The least scratching of the finish is obtained by brush washing with ample supplies of water – the traditional hose and brush, of which one of the best available is the Gardena, incorporating a flow control tap in the back of the brush. Heavy deposits of mud should be hosed first, to ensure thorough wetting and to get rid of the excess grit. Brush-washing then follows from the roof downwards, and if the car-wash solvent which drops into the brush dispenser is used, there is no need to leather off. The solvent dries off without smears, and as well as being good for finish by avoiding scratching, this simple way of car washing saves effort to help compensate for extra water rates which may be incurred.

The washing process should include insides of door jambs and sills, underneath bumpers and front or rear aprons, and beneath sills. These out-of-sight areas, if neglected, become the weak points from which corrosion can begin.

Paintwork needs to be sealed by wax polishing, ideally to be done monthly; but taking a realistic view, the application of wax polish at least before and after the winter is a reasonable minimum if one wants paintwork to last well. This job should always be tackled on the same day as that on which the car has been washed; if any longer interval is allowed, dust will have settled on the paintwork, which is then ground in with the polish, forming an abrasive compound and causing scratching. Both when applying and removing the polish, it should be

139

applied with to-and-fro movement along the lines of the bodywork, and circular movement should be avoided. The way in which rainwater stands in globules on a newly wax-polished car shows how this protects the finish, and modern liquid wax polishes have now taken all the hard work out of the job of application.

Paintwork repairs

It is almost inevitable that paintwork should become damaged in service, varying from small chips caused by flying stones or someone else's carelessly opened door, to more substantial damage resulting from a minor graze or manoeuvring dent. If the surface paint has been broken, early attention is important, to prevent rust from taking a hold. The old technique of just touching-up minor blemishes of this kind with a touch-up brush is a simple but very crude way of repairing the damage, and as well as looking rather untidy such a repair will not last.

The most important thing to bear in mind about paint is that it is not a filler – it is just a colour. Brushed on paint over a chip will simply change its colour; the blemish will remain. With a little care and taking advantage of modern materials, the work can be much better done, and the damage completely concealed. The work falls into two stages – making good, and painting. If the damage is not treated almost as soon as it has happened, a third stage precedes these two – 'killing' the rust.

A small area surrounding the blemish must be rubbed down to the metal, the aim being to obtain a shiny surface of steel showing where the damage had been, tapering gradually up to an area of surrounding paintwork which is unaffected. The rubbing down must be done with wet-and-dry emery paper, coarse at first if the damage is substantial or if there is a lot of rust to remove, and then finishing with finer paper to eliminate scratches. A little rust inhibitor such as Jenolite is now applied following the maker's directions. The surface may now need filling, which is the second stage – or the first if no rust was present. Body filler is obtainable from accessory shops as a greyish semi-liquid compound; the required amount is mixed with the hardener and applied straight away using the plastic spreader usually supplied. The key to success here is not to mix too much compound at a time, and to make the application promptly after mixing, and do so with a firm definite move, much as one might dash some butter on to a piece of toast. Further attempts to spread it better and improve the surface usually finish by making it worse than it was, so it is a case of leaving well alone until it has firmly set, often taking an hour or so.

When the filler has become completely hard – checked by testing the remainder of the mixture rather than that on the damaged area – more may be mixed if necessary for further filling, or the surface may be rubbed smooth. The process of filling, rubbing down and filling again if required must continue until a perfectly smooth surface has been obtained, completing the stage of making good.

Painting is easily done with aerosol sprays, first a primer spray which must be allowed to dry thoroughly before lightly rubbing down with fine emery paper and water or cutting down paste and cloth, which again can be moistened to give a finer finish. Finally, and only when one is satisfied that the surface is perfectly smooth and that no break can be detected between the original surrounding paint and the filling and priming, the matching paint colour for the car is sprayed over the whole area.

Aerosol paint sprays give good results if used correctly, but a little practice is advisable first to make sure that the spray is flowing correctly and that the colour is right; so test on a piece of scrap metal or wood that does not matter, before attempting to spray the car.

141

For spraying a local area, a little overspray on to the surrounding undamaged area is permissible, and in due course the new and the old should merge into one. Nearby chrome, wheels or parts of different colour need to be masked using masking tape and newspaper. Do not spray right up to the masking, or a sharp line of new and old paint will result, which will always show. Most aerosol paints have to be shaken hard to disperse the paint solvents by an agitator ball, which can be heard rattling inside. Very thorough shaking, perhaps for as much as a minute or more, is essential to prepare a paint aerosol for use. The canister should not be exposed to excessive cold prior to use; it should be kept in a reasonably warm room rather than in the garage. Spraying is done with the canister at least a foot away from the surface to be painted, and with deliberate lateral movements, the jet being controlled by the release button at the top.

Common faults when spraying in this way are to apply with the canister too close and moving too slowly, so that excess paint forms, resulting in ugly runs; and spraying too many coats without allowing time to dry can produce the same effect. The best results are obtained by applying three or four coats at approximately $\frac{1}{4}$-hour intervals. The new paint should be left to harden for about a fortnight before attempting to polish down, using cutting paste on a damp cloth. Polishing too soon can result in tearing the surface or embedding the cutting paste into it. With care, surprisingly professional results can be obtained using these readily available and reasonably inexpensive materials.

The other kind of damage referred to earlier is that in which the paintwork had not broken, but simply a dent or caving-in of the metal. Great care is needed in attempting to push this out, which can often do more damage than was caused in the pushing-in blow originally! The help of a professional panel beater is worth while, as he may be able to make a repair without spoiling the original paintwork. Simply pushing the panel out from the other side may still leave creases in the metal at the sharp points of the bend, unless the work is done with skill.

Brightwork

Stainless steel is now in extensive use for the brightwork of cars, in such areas as hub plates and bumpers, offering the advantage that it will not corrode. Rust may form on it, but it will be only of a surface nature, which can be removed even with the sponge or leather while washing the car. The surface can easily be scratched, however, and care should be taken to avoid using abrasive polishes or dirty cleaning cloths on stainless steel.

Chromium plate is still in extensive use, though, and often needs careful protection if it is not to rust and develop ugly pockmarks of corrosion. A chrome cleaner *and* polish, preferably a wax polish, need to be applied to all chromium parts at fairly regular intervals if it is intended to keep the car looking in good shape to preserve its value. Once pitting in chromium has started it is very difficult to restrict it.

When wax polishing the bodywork it is a good idea to apply it liberally at the upper junction between trim strips and the paintwork, so that a waterproof seal is formed; but not at the bottom – you want water to be kept out not kept in.

Doors and lids

Although it is common practice nowadays for a rubber seal to bear against the outer surface of the window at its lower edge, water still finds its way down inside the door in rain. Under the lower edge of the door will be found one or more little outlets for this water to escape, and a periodic check should be

143

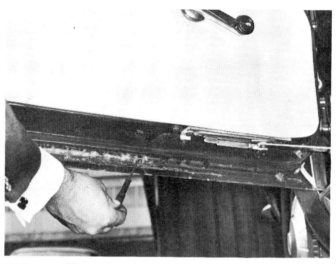

Using a small screwdriver to check that door drain holes are not blocked on a Rover 2000. If silt and dirt are allowed to accumulate and trap the exit, water will cause rapid corrosion.

made to ensure that these drainways are kept clear. A thin knitting-needle is a good tool for the purpose. If the drainways are allowed to block with silt and rust, water collects in the bottom of the door, causing rapid rust from the inside, usually at the point where the outer skin forms a flange with the inner pressing.

The frequent prowl round the car with the oil can should include a drop or two of oil at the tops of push button releases, pivots of door keeps, and on the top of hinges of doors, boot lid and bonnet. A glance at the locking mechanism will show where the striker engages, and here a smear of grease is the best treatment. Excess should be wiped off to prevent grease stains on clothing of people getting into and out of the car.

144

Maladjustment of door hinges makes itself evident in the form of bright metal showing through the worn-away paint at the point where the two metal surfaces are fouling each time the door is closed, and by the grating sound and feel of the door as it is closed. Usually there is adjustment both at the striker plates of the lock and at the hinges. Try to rectify by adjusting the latch before resorting to the hinges. The correct screwdriver – normally Philips cross-head type – must be used, to avoid unsightly damage to the screws. One hinge should be slackened well, and the other just loosened; the door is then positioned for smooth closing, re-opened while taking the weight off the hinges so that the adjustment is not lost, and the screws re-tightened. If a fit is not achieved, the first hinge should be tightened-up slightly and the other loosened. Use a flat piece of wood to check that the front edge of the door does not stand clear of the adjoining wing, as this will lead to excessive wind noise. When correct adjustment has been obtained, re-tighten really hard, otherwise the door will slip again in service.

Interior valeting

It is not intended to give detailed advice on anything so elementary as how to clean the inside of a car, but just to emphasize some basic principles. First, that if carpets are removable without a lot of unscrewing of fitments, they should be taken out and cleaned with an ordinary domestic vacuum cleaner, or beaten and brushed. Otherwise a plug-in 12-volt vacuum cleaner is worth while. By getting the ingrained dirt out of them, their life is extended. A light solution of warm water and a proprietary cleaner such as 1001 can be used to remove stains – check first for colour fastness in an area of carpet under the seats, which will not show if it proves that the cleaner is taking the colour out and leaving stains on the cleaner cloth.

Cleaning inside the engine compartment should not be forgotten: a clean engine is much more pleasant to work on, and adds to resale value. A solvent such as 'Gunk' or 'Jizer' should be sloshed on and spread over oil accumulations. An old washing-up mop is an ideal implement for the job. Then disperse it with liberal use of water, but take care not to soak the distributor.

A soft leather with mildly soapy water is the best treatment for removing the blue haze which forms on the inside windows of even a non-smoker's car, and instrument glasses should be dealt with at the same time. Seats, door trim, facia, sun visors and roof lining should be cleaned with liberal application of a proprietary upholstery cleaner. It is the accumulation of dust, stains and dirt in old, neglected cars which often upset those who travel in them.

6 · Seasonal Service

EVERY manufacturer lays down the recommended service procedure for his cars, based on research and actual experience of the life of components and the lubrication or adjustment needed. The best advice possible is that the official schedule should be followed at all times. Attention less often than is recommended may result in a less safe vehicle, not to mention a less efficient one, wasting fuel, wearing needlessly rapidly and giving below-standard performance. More often than recommended can also be a fault, though a much less serious one; and in fact it can be advisable to carry out such service procedures as the changing of engine oil and filter, in cars which run an exceptionally low annual mileage, at much shorter mileage intervals.

Because the manufacturer presumes that his car will be submitted to a service station for maintenance, the recommended procedures tend to be grouped together for attention at 5,000- or 6,000-mile intervals. This can result in a list involving too much work to be tackled at one time by the man who will service the car himself. It can also result in a lot of work to be done in a most unseasonal period, when a big maintenance schedule falls due in January or February.

To help the home mechanic, therefore, the following schedule is devised with the idea of spreading the load of car maintenance through the year, ensuring that certain jobs of a protective nature are done at the right time of the year, and that the work to be carried out is reduced in the heart of winter and at holiday time. However, it is not intended to over-ride the manufacturer's schedule, which should be checked to ensure that all service attentions demanded for your particular model are covered, and that the time intervals suggested will not – in your particular case – result in a higher mileage interval than the recommended one. If your annual mileage is such that in 4 months you may have covered 8,000 miles, then this interval is too great for an oil change on most cars, and the schedule should be modified accordingly. But because only 2,000 miles may have been covered, this does not mean that the interval can be extended: after 4 months' use, engine oil is generally due for a change.

The schedule is based on an annual mileage of 10,000 to 12,000. This is generally regarded as the limit of life for sparking plugs and contact points, so if the total is exceeded, these replacements should be brought forward as well.

For convenience, the service items are divided into two parts, with the idea that one stage of the month's work is done on one weekend, and the other half later during the same month. On each occasion the work is to be preceded by the routine checks at least once every 2 weeks. This includes a check of tyre pressures, but does not preclude combining this with re-fuelling, as many people like to do. Naturally, cleaning and valeting, including the important cleaning of headlamp glasses and interior surfaces of windows, are not separately specified. Now on to the schedule; the work advised should not – on average – take more than about a couple of hours, twice a month, a total of 48 hours in the year.

Fortnightly maintenance

These checks, numbered 1 to 7, should precede the appropriate part of the schedule for the month. The conscientious owner may like to run through the fortnightly schedule every week as well.

Check:	*Ref. page*
1 Tyre pressures.	21
2 Tyres for cuts and adequate tread depth.	15, 17
3 Brake fluid level.	45
4 Oil level.	62
5 Battery acid level.	57
6 Radiator (cold).	66
7 Windscreen washer reservoir.	59

Monthly maintenance

The fortnightly schedule is now followed by the appropriate half of the schedule for the month. In each case, the jobs number on from 7.

January

First half	*Ref. page*
1 to 7 Routine checks	
8 Check all lights	54
9 Check level in clutch hydraulic reservoir	119
10 Clean crank-case ventilator valve	105

Second half

1 to 7 Routine checks	
8 Fit new air filter element	103
9 Check that air intake is in correct position for winter running	104
10 Wax-polish car	139

February

First half	*Ref. page*
1 to 7 Routine checks	
8 Top-up carburettor dashpots	96
9 Lubricate all latches and hinges, throttle connections and (if provided) dynamo rear bearing; clear door drain holes	144 56
10 Inspect fan belt and adjust if necessary	56

Second half

4 to 7 Routine checks	
8 Change wheels round, combining with	19
9 Remove flints and debris from tyre treads, inspect inner tyre walls, and	16
10 Inspect front brakes; renew pads or linings if necessary; adjust, if drum brakes	28
11 Inspect rear brakes; renew pads or linings if necessary; adjust, if drum brakes	28
12 Check brake-fluid level	45
13 Adjust tyre pressures	21

March

First half	*Ref. page*
1 to 3 and 5 to 7 Routine checks	
8 Adjust tappets	107
9 Run car to check for leaks after re-fitting rocker cover, and to heat engine through; then drain oil	61
10 Fit new oil filter	62
11 Replace sump plug and pour in new oil	62
12 De-grease and clean engine; check for oil leaks and tightness of all fittings	146

Second half

1 to 7 Routine checks	
8 Fit new sparking plugs	69
9 Fit new contact points in distributor	76
10 Re-set ignition timing	81
11 Check oil level in steering box, or grease as required	50
12 Attend to any other grease points on car	49, 128

April

May

First half   *Ref. page*

1 to 7 Routine checks

8 Change oil in gearbox (and overdrive if fitted); if   126, 131
  no provision, check level

9 Change final drive oil in cars with live back axle; if   129
  no provision, check level

Second half

1 to 7 Routine checks

8 Have front wheel alignment and propeller shaft   8, 51, 136
  joints or drive shaft joints checked

9 Wax-polish car   139

June

First half	*Ref. page*
4 to 7 Routine checks	
8 Change wheels round, combining with	19
9 Remove flints and debris from tyre treads, inspect inner tyre walls, and	16
10 Inspect front brakes; renew pads or linings if necessary; adjust, if drum brakes	28
11 Inspect rear brakes; renew pads or linings if necessary; adjust, if drum brakes	28
12 Check brake-fluid level	45
13 Adjust tyre pressures	21

Second half

1 to 7 Routine checks	
8 Check oil level in steering box, or grease as required	50
9 Attend to any other grease points on car	49, 128
10 Check all steering and suspension connections	49
11 Check all flexible brake pipes	45

July

First half	*Ref. page*
1 to 3 and 5 to 7 Routine checks	
8 Change engine oil	61
9 Lubricate distributor	80
10 Lubricate all latches and hinges, throttle connections and (if provided) dynamo bearing	144 56

Second half

1 to 7 Routine checks	
8 Inspect fan belt and adjust if necessary; renew belt at least every third year	64
9 Clean crankcase ventilator valve	105
10 Check engine for oil or water leaks and tightness of all fittings; degrease if needed	146

August

First half	*Ref. page*
1 to 7 Routine checks	
8 Top-up carburettor dashpots	96
9 Check level in clutch reservoir	119

Second half

1 to 7 Routine checks	
8 Adjust tappets	107
9 Check all door drain holes	143
10 Wax-polish car	139

September

First half	*Ref. page*
1 to 7 Routine checks	
8 Fit new windscreen wiper blades (or new rubber inserts)	58
9 Check all lights and adjustment of headlamps	54
10 Check oil level in steering box, or grease as required	50
11 Attend to any other grease points on car	49, 128

Second half

1 to 7 Routine checks	
8 De-grease and clean engine; check for oil leaks and tightness of all fittings	146
9 Clean sparking plugs and check gaps	69
10 Check distributor points gap; re-adjust if necessary	79
11 Check ignition timing; re-adjust if necessary	81

October

First half	*Ref. page*

1 to 5 and 7 Routine checks

8 Check condition of water hoses and tightness of clips; renew as required	66
9 If suspect, check thermostat; renew if faulty	67
10 Drain off old anti-freeze; flush out thoroughly and fill with new anti-freeze solution	66
11 Lubricate distributor	80
12 Lubricate all latches and hinges, throttle connections and (if provided) dynamo bearing	144 56

Second half

4 to 7 Routine checks

8 Change wheels round, combining with	19
9 Remove flints and debris from tyre treads, inspect inner tyre walls, and	16
10 Inspect front brakes; renew pads or linings if necessary; adjust, if drum brakes	28
11 Inspect rear brakes; renew pads or linings if necessary; adjust, if drum brakes	28
12 Check brake fluid level	45
13 Check adjustment of handbrake; lubricate if required	35
14 Adjust tyre pressures	21

November

December

	Ref. page
(Only one service to allow for Christmas, but please repeat Routine Checks)	
1, 2 and 4 to 7 Routine checks	
8 Check all steering and suspension connections	49
9 Check oil-level in steering box, or grease as required	50
10 Attend to any other grease points on car	49, 128
11 Check engine for oil or water leaks and tightness of all fittings; degrease if needed	146
12 Renew brake fluid by bleeding all round	46
13 In every third year, arrange for all flexible brake pipes and cylinder seals to be renewed, which will include item 12	45